Seed Production and Quality Control

Seed Production and Quality Control

– Author –
M.L. Alam

Wordpen Academics
Publishing Science That Matter

ISBN: 978-93-90438-67-9 (Hardback)

Published by : **WORDPEN ACADEMICS**
4736/23, Ansari Road, Darya Ganj,
New Delhi – 110 002
Phone: +91-011-35004336
E-mail: info@wordpenacademics.com

Printed at **:** **Replika Press Pvt. Ltd.**

Exclusive Distributor: ***Bio-Green Books, New Delhi***

Preface

A seed is a small embryonic plant enclosed in a covering called the seed coat, usually with some stored food. It is the product of the ripened ovule of gymnosperm and angiosperm plants which occurs after fertilization and some growth within the mother plant. The formation of the seed completes the process of reproduction in seed plants, with the embryo developed from the zygote and the seed coat from the integuments of the ovule. All seeds are different size, shape and colour. Production of genetically pure and otherwise good quality pedigree seed is an exacting task requiring high technical skills and comparatively heavy financial investment. During seed production strict attention must be given to the maintenance of genetic purity and other qualities of seeds in order to exploit the full dividends sought to be obtained by introduction of new superior crop plant varieties. In other words, seed production must be carried out under standardized and well-organized condition.

Seeds are produced in several related groups of plants, and their manner of production distinguishes the angiosperms ("enclosed seeds") from the gymnosperms ("naked seeds"). Angiosperm seeds are produced in a hard or fleshy structure called a fruit that encloses the seeds, hence the name. (Some fruits have layers of both hard and fleshy material). In gymnosperms, no special structure develops to enclose the seeds, which begin their development "naked" on the bracts of cones. However, the seeds do become covered by the cone scales as they develop in some species of conifer. Seed production in natural plant populations vary widely from year-to-year in response to weather variables, insects and diseases, and internal cycles within the plants themselves. Over a 20-year period, for example, forests composed of loblolly pine and shortleaf pine produced from 0 to nearly 5 million sound pine seeds per hectare. Over this period, there were six bumper, five poor, and nine good seed crops, when evaluated in regard to producing adequate seedlings for natural forest reproduction. Seeds are very diverse in size. The dust-like orchid seeds are the smallest, with about one million seeds per gram; they are often embryonic seeds with immature embryos and no significant energy reserves.

Orchids and a few other groups of plants are mycoheterotrophs which depend on mycorrhizal fungi for nutrition during germination and the early growth of the seedling. Some terrestrial orchid seedlings, in fact, spend the first few years of their lives deriving energy from the fungi and do not produce green leaves. At over 20 kg, the largest seed is the *coco de mer*. Plants that produce smaller seeds can generate many more seeds per flower, while plants with larger seeds invest more resources into those seeds and normally produce fewer seeds. Small seeds are quicker to ripen and can be dispersed sooner, so fall blooming plants often have small seeds. Many annual plants produce great quantities of smaller seeds; this helps to ensure at least a few will end in a favourable place for growth. Herbaceous perennials and woody plants often have larger seeds; they can produce seeds over many years, and larger seeds have more energy reserves for germination and seedling growth and produce larger, more established seedlings after germination. The seed industry in India has two distinct segments the Public sector seed organizations and the private sector seed agencies. The private seed sector includes private seed companies, seed dealers and seed producers. In India, public sector has not delivered the desired results for high quality seed production. Therefore, to ensure farmers a high quality seed from outside the country, i.e. anywhere in the world for which the Government of India announced in 1988, a new policy for seed development, based upon the recognition that hybrid seeds and hybrids of plantlets developed through tissue- culture are a high-technology and for which the participation of MRTP/FERA companies is a must. Accordingly seven MRTP and two FERA private industrial companies are engaged in manufacturing a high-quality hybrid seeds from imported parent material for use in Indian local conditions. Seed production is a highly technical practice which is done by scientists and is getting highly commercialized. However, the basic knowledge about the principles of seed production will help in people selecting and buying seeds as well as in the production of one's own seed materials.

This book incorporates articles on some of the very important aspects related to seed technology. This volume contains eight chapters covering different aspects related to seed production, quality control, seed processing, packaging, seed physiology and related issues. Major chapters included are on : Rules and regulations governing quality seed production, certification, seed law enforcement and penalties on contravenes; Role of quality seeds in nutritional security. This book will definitely serve as an excellent reference material and practical guide for students of seed technology, Agriculture, Horticulture, Botany, Forestry and Life Sciences.

—Dr. M.A. Alam

Contents

1

Introduction

Seed

A seed is a small embryonic plant enclosed in a covering called the seed coat, usually with some stored food. It is the product of the ripened ovule of gymnosperm and angiosperm plants which occurs after fertilization and some growth within the mother plant. The formation of the seed completes the process of reproduction in seed plants (started with the development of flowers and pollination), with the embryo developed from the zygote and the seed coat from the integuments of the ovule. Seeds have been an important development in the reproduction and spread of flowering plants, relative to more primitive plants such as mosses, ferns and liverworts, which do not have seeds and use other means to propagate themselves. This can be seen by the success of seed plants (both gymnosperms and angiosperms) in dominating biological niches on land, from forests to grasslands both in hot and cold climates. The term "seed" also has a general meaning that antedates the above — anything that can be sown, e.g. "seed" potatoes, "seeds" of corn or sunflower "seeds". In the case of sunflower and corn "seeds", what is sown is the seed enclosed in a shell or husk, whereas the potato is a tuber.

Seed Structure

A typical seed includes three basic parts: (i) an embryo, (ii) a supply of nutrients for the embryo, and (3) a seed coat. The embryo is an immature plant from which a new plant will grow under proper conditions. The embryo has one cotyledon or seed leaf in monocotyledons, two cotyledons in almost all dicotyledons and two or more in gymnosperms. The radicle is the embryonic root. The plumule is the embryonic shoot. The embryonic stem above the point of attachment of the cotyledon(s) is the epicotyl. The embryonic stem below the point of attachment is the hypocotyl.

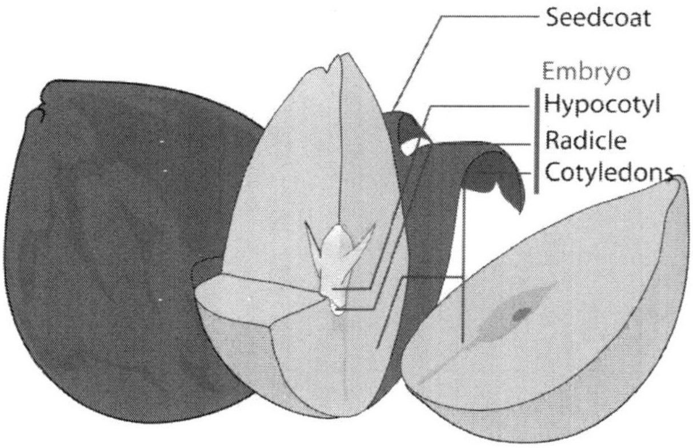

Figure: *The parts of an avocado seed (a dicot), showing the seed coat, endosperm, and embryo*

Within the seed, there usually is a store of nutrients for the seedling that will grow from the embryo. The form of the stored nutrition varies depending on the kind of plant. In angiosperms, the stored food begins as a tissue called the endosperm, which is derived from the parent plant via double fertilization. The usually triploid endosperm is rich in oil or starch and protein. In gymnosperms, such as conifers, the food storage tissue (also called endosperm) is part of the female gametophyte, a haploid tissue. In some species, the embryo is embedded in the endosperm or female gametophyte, which the seedling will use upon germination. In others, the endosperm is absorbed by the embryo as the latter grows within the developing seed, and the cotyledons of the embryo become filled with this stored food. At maturity, seeds of these species have no endosperm and are termed exalbuminous seeds. Some exalbuminous seeds are bean, pea, oak, walnut, squash, sunflower, and radish. Seeds with an endosperm at maturity are termed albuminous seeds. Most monocots (e.g. grasses and palms) and many dicots (e.g. Brazil nut and castor bean) have albuminous seeds. All gymnosperm seeds are albuminous.

The seed coat (the testa) develops from the tissue, the integument, originally surrounding the ovule. The seed coat in the mature seed can be a paper-thin layer (e.g. peanut) or something more substantial (e.g. thick and hard in honey locust and coconut). The seed coat helps protect the embryo from mechanical injury and from drying out. In addition to the three basic seed parts, some seeds have an appendage on the seed coat such an aril (as in yew and nutmeg) or an elaiosome (as in *Corydalis*) or hairs (as in cotton). A scar also may remain on the seed coat, called the hilum, where the seed was attached to the ovary wall by the funiculus.

Kinds of Seeds

Many structures commonly referred to as "seeds" are actually dry fruits. Sunflower seeds are sold commercially while still enclosed within the hard wall of the fruit, which must be split open to reach the seed. Different groups of plants have other modifications, the so-called stone fruits (such as the peach) have a hardened fruit layer (the endocarp) fused to and surrounding the actual seed. Nuts are the one-seeded, hard-shelled fruit of some plants with an indehiscent seed, such as an acorn or hazelnut.

Seed Production

Seeds are produced in several related groups of plants, and their manner of production distinguishes the angiosperms ("enclosed seeds") from the gymnosperms ("naked seeds"). Angiosperm seeds are produced in a hard or fleshy structure called a fruit that encloses the seeds, hence the name. (Some fruits have layers of both hard and fleshy material). In gymnosperms, no special structure develops to enclose the seeds, which begin their development "naked" on the bracts of cones. However, the seeds do become covered by the cone scales as they develop in some species of conifer.

Seed production in natural plant populations vary widely from year-to-year in response to weather variables, insects and diseases, and internal cycles within the plants themselves. Over a 20-year period, for example, forests composed of loblolly pine and shortleaf pine produced from 0 to nearly 5 million sound pine seeds per hectare. Over this period, there were six bumper, five poor, and nine good seed crops, when evaluated in regard to producing adequate seedlings for natural forest reproduction.

Seed Development

The seed, which is an embryo with two points of growth (one of which forms the stem, the other the roots) is enclosed in a seed coat with some food reserves. Angiosperm seeds consist of three genetically distinct constituents: (1) the embryo formed from the zygote, (2) the endosperm, which is normally triploid, (3) the seed coat from tissue derived from the maternal tissue of the ovule. In angiosperms, the process of seed development begins with double fertilization and involves the fusion of the egg and sperm nuclei into a zygote. The second part of this process is the fusion of the polar nuclei with a second sperm cell nucleus, thus forming a primary endosperm. Right after fertilization, the zygote is mostly inactive, but the primary endosperm divides rapidly to form the endosperm tissue. This tissue becomes the food the young plant will consume until the roots have developed after germination. The seed coat forms from the two integuments or outer layers of cells of the ovule, which derive from tissue from the mother plant, the inner integument forms the tegmen and the outer forms the testa. When the seed coat forms from only one layer, it is also called the testa, though not all such testae are homologous from one species to the next.

In gymnosperms, the two sperm cells transferred from the pollen do not develop seed by double fertilization, but one sperm nucleus unites with the egg nucleus and the other sperm is not used. Sometimes each sperm fertilizes an egg cell and one zygote is then aborted or absorbed during early development. The seed is composed of the embryo (the result of fertilization) and tissue from the mother plant, which also form a cone around the seed in coniferous plants such as pine and spruce.

The ovules after fertilization develop into the seeds; the main parts of the ovule are the funicle; which attaches the ovule to the placenta, the nucellus; the main region of the ovule where the megagametophyte develops, the micropyle; a small pore or opening in the ovule where the pollen tube usually enters during the process of fertilization, and the chalaza; the base of the ovule opposite the micropyle, where integument and nucellus are joined together.

The shape of the ovules as they develop often affects the final shape of the seeds. Plants generally produce ovules of four shapes: the most common shape is called anatropous, with a curved shape. Orthotropous ovules are straight with all the parts of the ovule lined up in a long row producing an uncurved seed. Campylotropous ovules have a curved megagametophyte often giving the seed a tight "C" shape. The last ovule shape is called amphitropous, where the ovule is partly inverted and turned back 90 degrees on its stalk (the funiculus).

In the majority of flowering plants, the zygote's first division is transversely oriented in regards to the long axis, and this establishes the polarity of the embryo. The upper or chalazal pole becomes the main area of growth of the embryo, while the lower or micropylar pole produces the stalk-like suspensor that attaches to the micropyle. The suspensor absorbs and manufacturers nutrients from the endosperm that are used during the embryo's growth.

The embryo is composed of different parts; the epicotyle will grow into the shoot, the radicle grows into the primary root, the hypocotyl connects the epicotyle and the radicle, the cotyledons form the seed leaves. Monocotyledonous plants have other structures; instead of the hypocotyle-epicotyle, it has a coleoptile that forms the first leaf and connects to the coleorhiza that connects to the primary root and adventitious roots form from the sides. The seeds of corn are constructed with these structures; pericarp, scutellum (single large cotyledon) that absorbs nutrients from the endosperm, endosperm, plumule, radicle, coleoptile and coleorhiza - these last two structures are sheath-like and enclose the plumule and radicle, acting as a protective covering. The testae or seed coats of both monocots and dicots are often marked with patterns and textured markings, or have wings or tufts of hair.

Seed Size and Seed Set

Seeds are very diverse in size. The dust-like orchid seeds are the smallest, with about one million seeds per gram; they are often embryonic seeds with immature embryos and no significant energy reserves. Orchids and a few other groups of plants are mycoheterotrophs which depend on mycorrhizal fungi for nutrition during germination and the early growth of the seedling. Some terrestrial orchid seedlings, in fact, spend the first few years of their lives deriving energy from the fungi and do not produce green leaves. At over 20 kg, the largest seed is the *coco de mer*. Plants that produce smaller seeds can generate many more seeds per flower, while plants with larger seeds invest more resources into those seeds and normally produce fewer seeds. Small seeds are quicker to ripen and can be dispersed sooner, so fall blooming plants often have small seeds. Many annual plants produce great quantities of smaller seeds; this helps to ensure at least a few will end in a favourable place for growth. Herbaceous perennials and woody plants often have larger seeds; they can produce seeds over many years, and larger seeds have more energy reserves for germination and seedling growth and produce larger, more established seedlings after germination.

Seed Functions

Seeds serve several functions for the plants that produce them. Key among these functions are nourishment of the embryo, dispersal to a new location, and dormancy during unfavourable conditions. Seeds fundamentally are means of reproduction, and most seeds are the product of sexual reproduction which produces a remixing of genetic material and phenotype variability on which natural selection acts.

Embryo Nourishment

Seeds protect and nourish the embryo or young plant. They usually give a seedling a faster start than a sporeling from a spore, because of the larger food reserves in the seed and the multicellularity of the enclosed embryo.

Seed Dispersal

Seed dispersal is the movement or transport of seeds away from the parent plant. Plants have limited mobility and consequently rely upon a variety of dispersal vectors to transport their propagules, including both abiotic and biotic vectors. Seeds can be dispersed away from the parent plant individually or collectively, as well as dispersed in both space and time. The patterns of seed dispersal are determined in large part by the dispersal mechanism and this has important implications for the demographic and genetic structure of plant populations, as well as migration patterns and species interactions. There are five main modes of seed dispersal: gravity, wind, ballistic, water and by animals. Some plants are serotinous and only disperse their seeds in response to an environmental stimulus. It can be measured using seed traps.

Benefits of Seed Dispersal

Seed dispersal is likely to have several benefits for plant species. First, seed survival is often higher away from the parent plant. This higher survival may result from the actions of density-dependent seed and seedling predators and pathogens, which often target the high concentrations of seeds beneath adults. Competition with adult plants may also be lower when seeds are transported away from their parent. Seed dispersal also allows plants to reach specific habitats that are favourable for survival, a hypothesis known as directed dispersal. For example, *Ocotea endresiana* (Lauraceae) is a tree species from Latin America which is dispersed by several species of birds, including the three-wattled bellbird. Male bellbirds perch on dead trees in order to attract mates, and often defecate seeds beneath these perches where the seeds have a high chance of survival because of high light conditions and escape from fungal pathogens. In the case of fleshy-fruited plants, seed-dispersal in animal guts (endozoochory) often enhances the amount, the speed, and the asynchrony of germination, which can have important plant benefits. Seeds dispersed by ants (myrmecochory) are not only dispersed to short distances but are also buried underground by the ants. These seeds can thus avoid adverse environmental effects such as fire or drought, reach nutrient-rich microsites and survive longer than other seeds. These features are peculiar to myrmecochory, which may thus provide additional benefits not present in other dispersal modes. Finally, at another scale, seed dispersal may allow plants to colonize vacant habitats and even new geographic regions.

Types of Dispersal

Seed dispersal is sometimes split into autochory (when dispersal is attained using the plant's own means) and allochory (when obtained through external means).

Autochory

Gravity: Barochory or the plant use of gravity for dispersal is a simple means of achieving seed dispersal. The effect of gravity on heavier fruits causes them to fall from the plant when ripe. Fruits exhibiting this type of dispersal include apples, coconuts and passionfruit and those with harder shells (which often roll away from the plant to gain more distance). Gravity dispersal also allows for later transmission by water or animal. Other types of autochory are boleochory (the seed is forcefully ejected by dehiscence and squeezing) or herpochory (the seed crawls by means of trichomes and changes humidity).

Allochory

Wind: ind dispersal (anemochory) is one of the more primitive means of dispersal. Wind dispersal can take on one of two primary forms: seeds can float on the breeze or alternatively, they can flutter to the ground. The classic examples of these dispersal mechanisms include dandelions, which have a feathery pappus attached to their seeds and can be dispersed long distances, and maples, which have winged seeds (samara) and flutter to the ground. An important constraint on wind dispersal is the need for abundant seed production to maximise the likelihood of a seed landing in a site suitable for germination. There are also strong evolutionary constraints on this dispersal mechanism. For instance, Cody and Overton (1996) found that species in the Asteraceae on islands tended to have reduced dispersal capabilities (i.e., larger seed mass and smaller pappus) relative to the same species on the mainland. Reliance on wind dispersal is common among many weedy or ruderal species. Unusual mechanisms of wind dispersal include tumbleweeds.

Water: Many aquatic (water) and some terrestrial (ground) plant species use hydrochory, or seed dispersal through water. Seeds can travel for extremely long distances, depending on the specific mode of water dispersal.

The water lily is an example of such a plant. Water lilies' flowers make a fruit that floats in the water for a while and then drops down to the bottom to take root on the floor of the pond. The seeds of palm trees can also be dispersed by water. If they grow near oceans, the seeds can be transported by ocean currents over long distances, allowing the seeds to be dispersed as far as other continents. Mangrove trees live right in the water. Their seeds fall from the tree and grow roots as soon as they touch any kind of soil. During low tide, they might fall in soil instead of water and start growing right where they fell. If the water level is high, however, they can be carried far away from where they fell. Mangrove trees often make little islands as dirt and other things collect in their roots, making little bodies of land. An special review for oceanic waters hydrochory can be seen at oceanic dispersal.

Dispersal by Animals: Animals can disperse plant seeds in several ways, all named zoochory. Seeds can be transported on the outside of vertebrate animals (mostly mammals), a process known as epizoochory. Plant species transported externally by animals can have a variety of adaptations for dispersal, including adhesive mucus, and a variety of hooks, spines and barbs. A typical example of an epizoochorous plant is *Trifolium angustifolium*, a species of Old World clover which adheres to animal fur by means of stiff hairs covering the seed. Epizoochorous plants tend to be herbaceous plants, with many representative species in the families Apiaceae and Asteraceae. However, epizoochory is a relatively rare dispersal syndrome for plants as a whole; the percentage of plant species with seeds adapted for transport on the outside of animals is estimated to be below 5%. Nevertheless, epizoochorous transport can be highly effective if seeds attach to wide-ranging animals. This form of seed dispersal has been implicated in rapid plant migration and the spread of invasive species.

Seed dispersal via ingestion by vertebrate animals (mostly birds and mammals), or endozoochory, is the dispersal mechanism for most tree species. Endozoochory is generally a coevolved mutualistic relationship in which a plant surrounds seeds with an edible, nutritious fruit as a good food for animals that consume it. Birds and mammals are the most important seed dispersers, but a wide variety of other animals, including turtles and fish, can transport viable seeds. The exact percentage of tree species dispersed by endozoochory varies between habitats, but can range to over 90% in some tropical rainforests. Seed dispersal by animals in tropical rainforests has received much attention, and this interaction is considered an important force shaping the ecology and evolution of vertebrate and tree populations. In the tropics, large animal seed dispersers (such as tapirs, chimpanzees and hornbills) may disperse large seeds with few other seed dispersal agents. The extinction of these large frugivores from poaching and habitat loss may have negative effects on the tree populations that depend on them for seed dispersal.

Seed dispersal by ants (myrmecochory) is a dispersal mechanism of many shrubs of the southern hemisphere or understorey herbs of the northern hemisphere. Seeds of myrmecochorous plants have a lipid-rich attachment called the elaiosome, which attracts ants. Ants carry such seeds into their colonies, feed the elaiosome to their larvae and discard the otherwise intact seed in an underground chamber. Myrmecochory is thus a coevolved mutualistic relationship between plants and seed-disperser ants. Myrmecochory has independently evolved at least 100 times in flowering plants and is estimated to be present in at least 11 000 species, but likely up to 23 000 or 9% of all species of flowering plants. Myrmecochorous plants are most frequent in the fynbos vegetation of the Cape Floristic Region of South Africa, the kwongan vegetation and other dry habitat types of Australia, dry forests and grasslands of the Mediterranean region and northern temperate forests of western Eurasia and eastern North America, where up to 30-40% of understorey herbs are myrmecochorous. Seed predators, which include many rodents (such as squirrels) and some birds (such as jays) may also disperse seeds by hoarding the seeds in hidden caches. The seeds in caches are usually well-protected from other seed predators and if left uneaten will grow into new plants. In addition, rodents may also disperse seeds via seed spitting due to the presence of secondary metabolites in ripe fruits. Finally, seeds may be secondarily dispersed from seeds deposited by primary animal dispersers. For example, dung beetles are known to disperse seeds from clumps of feces in the process of collecting dung to feed their larvae.

Other types of zoochory are chiropterochory (by bats), malacochory (by molluscs, mainly terrestrial snails) or ornithochory (by birds).

Dispersal by Humans: Dispersal by humans (anthropochory) used to be seen as a form of dispersal by animals. Recent research points out that human dispersers differ from animal dispersers by a much higher mobility based on the technical means of human transport. Dispersal by humans on the one hand may act on large geographical scales and lead to invasive species. On the other hand dispersal by humans also acts on smaller, regional scales and drive the dynamics of existing biological populations. Humans may disperse seeds by many various means and some surprisingly high distances have been repeatedly measured. Examples are: dispersal on human clothes (up to 250m), on shoes (up to 5 km) or by cars (regularly ~ 250m, singles cases > 100 km).

Consequences of Seed Dispersal

Seed dispersal has many consequences for the ecology and evolution of plants. Dispersal is necessary for species migrations, and in recent times dispersal ability is an important factor in whether or not a species transported to a new habitat by humans will become an invasive species. Dispersal is also predicted to play a major role in the origin and maintenance of species diversity. For example, myrmecochory increased the rate of diversification more than twofold in plant groups in which it has evolved because myrmecochorous lineages contain more than twice as many species as their non-myrmecochorous sister groups. Dispersal of seeds away from the parent organism has a central role in two major theories for how biodiversity is maintained in natural ecosystems, the Janzen-Connell hypothesis and recruitment limitation. Seed dispersal is essential in allowing forest migration.

Seed Dormancy

Seed dormancy is a condition of plant seeds that prevents germination when the seeds are under optimal environmental conditions for germination. Living, non dormant seeds germinate when soil temperatures and moisture conditions are suited for cellular processes and division; dormant seeds do not. One important function of most seeds is delayed germination, which allows time for dispersal and prevents germination of all the seeds at same time. The staggering of germination safeguards some seeds and seedlings from suffering damage or death from short periods of bad weather or from transient herbivores; it also allows some seeds to germinate when competition from other plants for light and water might be less intense. Another form of delayed seed germination is seed quiescence, which is different than true seed dormancy and occurs when a seed fails to germinate because the external environmental conditions are too dry or warm or cold for germination. Many species of plants have seeds that delay germination for many months or years, and some seeds can remain in the soil seed bank for more than 50 years before germination. Some seeds have a very long viability period, and the oldest documented germinating seed was nearly 2000 years old based on radiocarbon dating.

Overview

True dormancy or innate dormancy is caused by conditions within the seed that prevent germination under normally ideal conditions. Often seed dormancy is divided into two major categories based on what part of the seed produces dormancy: exogenous and endogenous. There are three types of dormancy based on their mode of action: physical, physiological and morphological.

There have been a number of classification schemes developed to group different dormant seeds, but none have gained universal usage. Dormancy occurs because of a wide range of reasons that often overlap, producing conditions in which definitive categorization is not clear. Compounding this problem is that the same seed that is dormant for one reason at a given point may be dormant because of another reason at a later point. Some seeds fluctuate from periods of dormancy to non dormancy, and despite the fact that a dormant seed appears to be static or inert, in reality they are still receiving and responding to environmental cues.

Exogenous Dormancy

Exogenous dormancy is caused by conditions outside the embryo and is often broken down into three subgroups:

Physical Dormancy

Which occurs when seeds are impermeable to water or the exchange of gases. Legumes are typical examples of physically dormant seeds; they have low moisture content and are prevented from imbibing water by the seed coat. Chipping or cracking of the seed coat or any other coverings allows water intake. Impermeability is often caused by an outer cell layer which is composed of macrosclereid cells or the outer layer is composed of a mucilaginous cell layer. The third cause of seed coat impermeability is a hardened endocarp. Seed coats that are impermeable to water and gases form during the last stages of seed development.

Mechanical Dormancy

Mechanical dormancy occurs when seed coats or other coverings are too hard to allow the embryo to expand during germination. In the past this mechanism of dormancy was ascribed to a number of species that have been found to have endogenous factors for their dormancy instead. These endogenous facts include physiologically dormancy cased by low embryo growth potential

Chemical Dormancy

Includes growth regulators etc., that are present in the coverings around the embryo. They may be leached out of the tissues by washing or soaking the seed, or deactivated by other means. Other chemicals that prevent germination are washed out of the seeds by rainwater or snow melt.

Endogenous Dormancy

Endogenous dormancy is caused by conditions within the embryo itself, and it is also often broken down into three subgroups: physiological dormancy, morphological dormancy and combined dormancy, each of these groups may also have subgroups.

Physiological Dormancy

Physiological dormancy prevents embryo growth and seed germination until chemical changes occur. These chemicals include inhibitors that often retard embryo growth to the point where it is not strong enough to break through the seed coat or other tissues. Physiological dormancy is indicated when an increase in germination rate occurs after an application of gibberellic acid (GA3) or after Dry after-ripening or dry storage. It is also indicated when dormant seed embryos are excised and produce healthy seedlings: or when up to 3 months of cold (0–10°C) or warm (=15°C) stratification increases germination: or when dry after-ripening shortens the cold stratification period required. In some seeds physiological dormancy is indicated when scarification increases germination.

Physiological dormancy is broken when inhibiting chemicals are broken down or are no longer produced by the seed; often by a period of cool moist conditions, normally below (+4C) 39F, or in the case of many species in *Ranunculaceae* and a few others, (“5C) 24F. Abscisic acid is usually the growth inhibitor in seeds and its production can be affected by light. Some plants like Peony species have multiple types of physiological dormancy, one affects radicle (root) growth while the other affects plumule (shoot) growth. Seeds with physiological dormancy most often do not germinate even after the seed coat or other structures that interfere with embryo growth are removed. Conditions that affect physiological dormancy of seeds include:

- Drying; some plants including a number of grasses and those from seasonally arid regions need a period of drying before they will germinate, the seeds are released but need to have a lower moisture content before germination can begin. If the seeds remain moist after dispersal, germination can be delayed for many months or even years. Many herbaceous plants from temperate climate zones have physiological dormancy that disappears with drying of the seeds. Other species will germinate after dispersal only under very narrow temperature ranges, but as the seeds dry they are able to germinate over a wider temperature range.
- Photodormancy or light sensitivity affects germination of some seeds. These photoblastic seeds need a period of darkness or light to germinate. In species with thin seed coats, light may be able to penetrate into the dormant embryo. The presence of light or the absence of light may trigger the germination process, inhibiting germination in some seeds buried too deeply or in others not buried in the soil.
- Thermodormancy is seed sensitivity to heat or cold. Some seeds including cocklebur and amaranth germinate only at high temperatures (30C or 86F) many plants that have seed that germinate in early to mid summer have thermodormancy and germinate only when the soil temperature is warm. Other seeds need cool soils to germinate, while others like celery are inhibited when soil temperatures are too warm. Often thermodormancy requirements disappear as the seed ages or dries.

Seeds are classified as having deep physiological dormancy under these conditions: applications of GA3 does not increase germination; or when excised embryos produce abnormal seedlings; or when seeds require more than 3 months of cold stratification to germinate.

Morphological Dormancy

Embryo underdeveloped or undifferentiated. Some seeds have fully differentiated embryos that need to grow more before seed germination, or the embryos are not differentiated into different tissues at the time of fruit ripening.

- Immature embryos – some plants release their seeds before the tissues of the embryos have fully differentiated, and the seeds ripen after they take in water while on the ground, germination can be delayed from a few weeks to a few months.

Combined Dormancy

Seeds have both morphological and physiological dormancy.

- Morpho-physiological or morphophysiological dormancy occurs when seeds with underdeveloped embryos, also have physiological components to dormancy. These seeds therefore require dormancy-breaking treatments as well as a period of time to develop fully grown embryos.

- Intermediate simple
- Deep simple
- Deep simple epicotyl
- Deep simple double
- Intermediate complex
- Deep complex

Combinational Dormancy

Combinational dormancy occurs in some seeds, where dormancy is caused by both exogenous (physical) and endogenous (physiological) conditions, some *Iris* species have both hard impermeable seeds coats and physiological dormancy.

Secondary Dormancy

Secondary dormancy occurs in some non-dormant and post dormant seeds that are exposed to conditions that are not favourable for germination, like high temperatures. It is caused by conditions that occur after the seed has been dispersed. The mechanisms of secondary dormancy are not yet fully understood but might involve the loss of sensitivity in receptors in the plasma membrane.

Not all seeds undergo a period of dormancy, many species of plants release their seeds late in the year when the soil temperature is too low for germination or when the environment is dry. If these seeds are collected and sown in an environment that is warm enough, and/or moist enough, they will germinate. Under natural conditions non dormant seeds released late in the growing season wait until spring when the soil temperature rises or in the case of seeds dispersed during dry periods until it rains and there is enough soil moisture.

Seeds that do not germinate because they have fleshy fruits that retard germination are quiescent, not dormant. Many garden plants have seeds that will germinate readily as soon as they have water and are warm enough, though their wild ancestors had dormancy. These cultivated plants lack seed dormancy because of generations of selective pressure by plant breeders and gardeners that grew and kept plants that lacked dormancy. Seeds of some mangroves are viviparous and begin to germinate while still attached to the parent; they produce a large, heavy root, which allows the seed to penetrate into the ground when it falls.

Seed Hibernation

Seed hibernation is different from seed dormancy. While seed dormancy can be defined as "a seed not germinating when conditions, e.g. light, water/ nutrient availability or the presence of activating substances like smoke, are favourable", hibernation is the ability of a seed to remain in hibernation when there is a lack of things essential to its development (water, sunlight, nutrients, etc.) or in harsh conditions (extreme cold, extreme heat, hard ground, etc.). There is no specific time limit in which a seed may hibernate; the oldest seeds that have been found to be viable are of the Date Palm (Phoenix dactylifera): several seeds found in the Israeli fortress of Masada were radiocarbon dated to be 2000 years old. One seed was still able to germinate (Sallon et al., 2008 Science). When conditions are right for the particular type of seed, they can come out of hibernation and grow. Seeds display hibernation for very short periods of time every year. In autumn, seeds produced in summer may have a chance to grow while the soil is still warm, but instead do not grow until spring. The seeds have chemicals encoded into them that will not allow them to germinate until the soil warms up again, a sign of spring. This is called *after ripening*. Seeds may also go into hibernation for other causes. Many pines have adapted to forest fires that destroy trees by creating pine cones that open in extreme heat (fire) and release seeds coated with a material that will break down only in extreme heat. The seeds hibernate until this happens. Seeds of the *Canna Lily*, and other plants, have a hard coating that must be broken down by contact with rough rock, soil, etc. before they begin to grow. Until the shell is broken and water reaches the seed, the seed remains dormant. When released, some seeds (especially those in a desert environment) must wait for rain before the seed is triggered out of hibernation and begins to grow. There are many other causes of dormancy. It is an unclear phenomenon; the limits of seed hibernation are still unknown, and many triggers have yet to be discovered.

Seedling

A seedling is a young plant sporophyte developing out of a plant embryo from a seed. Seedling development starts with germination of the seed. A typical young seedling consists of three main parts: the radicle (embryonic root), the hypocotyl (embryonic shoot), and the cotyledons (seed leaves). The two classes of flowering plants (angiosperms) are distinguished by their numbers of seed leaves: monocotyledons (monocots) have one blade-shaped cotyledon, whereas dicotyledons (dicots) possess two round cotyledons. Gymnosperms are more varied. For example, pine seedlings have up to eight cotyledons. The seedlings of some flowering plants have no cotyledons at all. These are said to be acotyledons.

Photomorphogenesis

In developmental biology, photomorphogenesis is light-mediated development. The photomorphogenesis of plants is often studied by using tightly-frequency-controlled light sources to grow the plants.

Germination

Light has profound effects on the development of plants. The light-mediated changes in plant growth and development are called photomorphogenesis. The most striking effects of light are observed when a germinating seedling emerges from the soil and is exposed to light for the first time.

Normally the seedling radicle (root) emerges first from the seed, and the shoot appears as the root becomes established. Later, with growth of the shoot (particularly when it merges into the light) there is increased secondary root formation and branching. This coordinated progression of developmental responses are early manifestations of correlative growth phenomena where the root affects the growth of the shoot and vice versa. To a large degree, these coordinated differential growth responses are hormone mediated.

In the absence of light, plants develop an etiolated growth pattern. Etiolation of the seedling adapts it to emerging from the soil.

Comparison of Dark-Grown (Etiolated) and Light-Grown (De-Etiolated) Seedlings

Etiolated characteristics:

- Distinct "apical hook" (dicot) or coleoptile (monocot)
- No leaf growth
- No chlorophyll
- Rapid stem elongation
- Limited radial expansion of stem
- Limited root elongation
- Limited production of lateral roots

De-etiolated characteristics:

- Apical hook opens or coleoptile splits open
- Leaf growth promoted
- Chlorophyll produced
- Stem elongation suppressed
- Radial expansion of stem
- Root elongation promoted
- Lateral root development accelerated

The developmental changes characteristic of photomorphogenesis shown by de-etiolated seedlings, are induced by light. Typically, plants are responsive to wavelengths of light in the blue, red and far-red regions of the spectrum through the action of several different photosensory systems. The photoreceptors for red and far-red wavelengths are known as phytochromes. There are at least 5 members of the phytochrome family of photoreceptors. There are several blue light photoreceptors.

Photoreceptor Systems in Plants

Plants use phytochrome to detect and respond to red and far-red wavelengths. Phytochromes are proteins with a light absorbing pigment attached (chromophore).

The chromophore is a linear tetrapyrrole called phytochromobilin. The phytochrome apoprotein is synthesized in the Pr form. Upon binding the chromophore, the holoprotein becomes sensitive to light. If it absorbs red light it will change conformation to the biologically active Pfr form. The Pfr form can absorb far red light and switch back to the Pr form.

Most plants have multiple phytochromes encoded by different genes. The different forms of phytochrome control different responses but there is also a lot of redundancy so that in the absence of one phytochrome, another may take on the missing functions.

Arabidopsis has 5 phytochromes - PHYA, PHYB, PHYC, PHYD, PHYE

Molecular analyses of phytochrome and phytochrome-like genes in higher plants (ferns, mosses, algae) and photosynthetic bacteria have shown that phytochromes evolved from prokaryotic photoreceptors that predated the origin of plants.

Blue Light Systems

As for the red/far-red system, plants contain multiple blue light photoreceptors which have different functions. Based on studies with action spectra, mutants and molecular analyses, it has been determined that higher plants contain at least 4, and probably 5, different blue light photoreceptors.

Cryptochromes were the first blue light receptors to be isolated and characterized from any organism. The proteins use a flavin as a chromophore. The cryptochromes have evolved from microbial DNA-photolyase, an enzyme that carries out light-dependent repair of UV damaged DNA.

Two cryptochromes have been identified in plants. Cryptochromes control stem elongation, leaf expansion, circadian rhythms and flowering time. In addition to blue light, cryptochromes also perceive long wavelength UV irradiation (UV-A). Phototropin is the blue light photoreceptor that controls phototropism. It also uses flavin as chromophore. Only one phototropin has been identified so far (NPH1). Phototropin also perceives long wavelength UV irradiation (UV-A) in addition to blue light.

Recent experiments indicate that a 4th blue light receptor exists that uses a carotenoid as a chromophore. This new photoreceptor controls blue light induction of stomatal opening. However, the gene and protein have not yet been found. Other blue light responses exist that seem to function in plants that are missing the cryptochrome, phototropin and carotenoid photoreceptors suggesting that at least one more will be found. Since the cryptochromes were discovered in plants, several labs have identified homologous genes and photoreceptors in a number of other organisms, including humans, mice and flies.

UV Systems

Based on various responses to UV light, it is assumed that there are UV-specific photoreceptor...

Etiolation

Figure: *Spanish Bluebells* Hyacinthoides hispanica, *showing both leaves and flowers in both etiolated and non-etiolated states. The longest etiolated leaves are about 50 cm long*

Etiolation (pronounced /iÐtjYÈlej*f*Yn/) is a process in flowering plants grown in partial or complete absence of light. It is characterized by long, weak stems; smaller, sparser leaves due to longer internodes; and a pale yellow colour (chlorosis).

Effects

Etiolation increases the likelihood that a plant will reach a light source, often from under the soil, leaf litter, or shade from competing plants. The growing tips are strongly attracted to light and will elongate towards it. The pale colour results from a lack of chlorophyll.

Some of the changes that occur include:

1. elongation of stems and leaves;
2. weakening of cell walls in stems and leaves;
3. longer internodes, hence fewer leaves per unit length of stem;
4. chlorosis, a pale yellowish-white colouration.

Causes

The cause of the elongation is due to auxins. Auxins are not active in light; when they are active they stimulate proton pumps in the cell wall which increases the acidity of the cell wall and activates expansin (an enzyme that breaks bonds in the cell wall structure) that weaken the cell wall and allow the cell to expand. Chloroplasts that have not been exposed to light are called etioplasts.

De-etiolation

De-etiolation, on the other hand, is a series of physiological and biochemical changes a plant shoot undergoes in response to sunlight. This process is also known informally as greening. The changes triggered in the plant shoot all occur in preparation for photosynthesis.

Some of the changes that occur include:

1. Inhibition of hypocotyl lengthening.
2. Stimulation of cotyledon expansion.
3. Opening of the apical hook.
4. Stimulation of synthesis of anthocyanins.
5. Stimulation of chloroplasts development from etioplasts.

This process is regulated by the exposure of various photoreceptor pigments to light. Phytochrome A and phytochrome B both respond to an increasing proportion of red light to far-red light which occurs when the shoot comes out into the open. Cryptochrome 1 responds to increasing amounts of blue light when the shoot reaches the surface.

2

Seedling Growth and Maturation

Once the seedling starts to photosynthesize, it is no longer dependent on the seed's energy reserves. The apical meristems start growing and give rise to the root and shoot. The first "true" leaves expand and can often be distinguished from the round cotyledons through their species-dependent distinct shapes. While the plant is growing and developing additional leaves, the cotyledons eventually senesce and fall off. Seedling growth is also affected by mechanical stimulation, such as by wind or other forms of physical contact, through a process called thigmomorphogenesis. Seedlings are particularly vulnerable to attack by pests and diseases, and can consequently experience high mortality rates. Pests and diseases which are especially damaging to seedlings include damping off, cutworms, slugs and snails.

Germination

Germination is the process by which plants, fungus and bacteria emerge from seeds and spores, and begin growth. The most common example of germination is the sprouting of a seedling from a seed of an angiosperm or gymnosperm. However the growth of a sporeling from a spore, for example the growth of hyphae from fungal spores, is also germination. In a more general sense, germination can imply anything expanding into greater being from a small existence or germ.

Seed Germination

Germination is the growth of an embryonic plant contained within a seed; it results in the formation of the seedling. The seed of a vascular plant is a small package produced in a fruit or cone after the union of male and female sex cells. All fully developed seeds contain an embryo and, in most plant species some store of food reserves, wrapped in a seed coat. Some plants produce varying numbers of seeds that lack embryos; these are called empty seeds and never germinate. Most seeds go through a period of dormancy where there is no active growth; during this time the seed can be safely transported to a new location and/or survive adverse climate conditions until circumstances are favourable for growth. Dormant seeds are ripe seeds that do not germinate because they are subject to external environmental conditions that prevent the initiation of metabolic processes and cell growth. Under favourable conditions, the seed begins to germinate and the embryonic tissues resume growth, developing towards a seedling.

Factors Affecting Seed Germination

Seed germination depends on both internal and external conditions. The most important external factors include temperature, water, oxygen and sometimes light or darkness. Various plants require different variables for successful seed germination. Often this depends on the individual seed variety and is closely linked to the ecological conditions of a plant's natural habitat. For some seeds, their future germination response is affected by environmental conditions during seed formation; most often these responses are types of seed dormancy.

- Water - is required for germination. Mature seeds are often extremely dry and need to take in significant amounts of water, relative to the dry weight of the seed, before cellular metabolism and growth can resume. Most seeds need enough water to moisten the seeds but not enough to soak them. The uptake of water by seeds is called imbibition, which leads to the swelling and the breaking of the seed coat. When seeds are formed, most plants store a food reserve with the seed, such as starch, proteins, or oils. This food reserve provides nourishment to the growing embryo. When the seed imbibes water, hydrolytic enzymes are activated which break down these stored food resources into metabolically useful chemicals. After the seedling emerges from the seed coat and starts growing roots and leaves, the seedling's food reserves are typically exhausted; at this point photosynthesis provides the energy needed for continued growth and the seedling now requires a continuous supply of water, nutrients, and light.
- Oxygen - is required by the germinating seed for metabolism. Oxygen is used in aerobic respiration, the main source of the seedling's energy until it grows leaves. Oxygen is an atmospheric gas that is found in soil pore spaces; if a seed is buried too deeply within the soil or the soil is waterlogged, the seed can be oxygen starved. Some seeds have

impermeable seed coats that prevent oxygen from entering the seed, causing a type of physical dormancy which is broken when the seed coat is worn away enough to allow gas exchange and water uptake from the environment.

- Temperature - affects cellular metabolic and growth rates. Seeds from different species and even seeds from the same plant germinate over a wide range of temperatures. Seeds often have a temperature range within which they will germinate, and they will not do so above or below this range. Many seeds germinate at temperatures slightly above room-temperature 60-75 F (16-24 C), while others germinate just above freezing and others germinate only in response to alternations in temperature between warm and cool. Some seeds germinate when the soil is cool 28-40 F (-2 - 4 C), and some when the soil is warm 76-90 F (24-32 C). Some seeds require exposure to cold temperatures (vernalization) to break dormancy. Seeds in a dormant state will not germinate even if conditions are favourable. Seeds that are dependent on temperature to end dormancy have a type of physiological dormancy. For example, seeds requiring the cold of winter are inhibited from germinating until they take in water in the fall and experience cooler temperatures. Four degrees Celsius is cool enough to end dormancy for most cool dormant seeds, but some groups, especially within the family Ranunculaceae and others, need conditions cooler than -5 C. Some seeds will only germinate after hot temperatures during a forest fire which cracks their seed coats; this is a type of physical dormancy.

Most common annual vegetables have optimal germination temperatures between 75-90 F (24-32 C), though many species (e.g. radishes or spinach) can germinate at significantly lower temperatures, as low as 40 F (4 C), thus allowing them to be grown from seed in cooler climates. Suboptimal temperatures lead to lower success rates and longer germination periods.

- Light or darkness - can be an environmental trigger for germination and is a type of physiological dormancy. Most seeds are not affected by light or darkness, but many seeds, including species found in forest settings, will not germinate until an opening in the canopy allows sufficient light for growth of the seedling.

Scarification mimics natural processes that weaken the seed coat before germination. In nature, some seeds require particular conditions to germinate, such as the heat of a fire (e.g., many Australian native plants), or soaking in a body of water for a long period of time. Others need to be passed through an animal's digestive tract to weaken the seed coat enough to allow the seedling to emerge.

Dormancy

Some live seeds are dormant and need more time, and/or need to be subjected to specific environmental conditions before they will germinate. Seed dormancy can originate in different parts of the seed, for example, within the embryo; in other cases the seed coat is involved. Dormancy breaking often involves changes in membranes, initiated by dormancy-breaking signals. This generally occurs only within hydrated seeds. Factors affecting seed dormancy include the presence of certain plant hormones, notably abscisic acid, which inhibits germination, and gibberellin, which ends seed dormancy. In brewing, barley seeds are treated with gibberellin to ensure uniform seed germination for the production of barley malt.

Seedling Establishment

In some definitions, the appearance of the radicle marks the end of germination and the beginning of "establishment", a period that ends when the seedling has exhausted the food reserves stored in the seed. Germination and establishment as an independent organism are critical phases in the life of a plant when they are the most vulnerable to injury, disease, and water stress. The germination index can be used as an indicator of phytotoxicity in soils. The mortality between dispersal of seeds and completion of establishment can be so high that many species have adapted to produce huge numbers of seeds.

Germination Rate

In agriculture and gardening, the germination rate describes how many seeds of a particular plant species, variety or seedlot are likely to germinate. It is usually expressed as a percentage, e.g., an 85% germination rate indicates that about 85 out of 100 seeds will probably germinate under proper conditions. The germination rate is useful for calculating the seed requirements for a given area or desired number of plants.

Dicot Germination

The part of the plant that first emerges from the seed is the embryonic root, termed the radicle or primary root. It allows the seedling to become anchored in the ground and start absorbing water. After the root absorbs water, an embryonic shoot emerges from the seed. This shoot comprises three main parts: the cotyledons (seed leaves), the section of shoot below the cotyledons (hypocotyl), and the section of shoot above the cotyledons (epicotyl). The way the shoot emerges differs among plant groups.

Epigeous

In epigeous (or epigeal) germination, the *hypocotyl* elongates and forms a hook, pulling rather than pushing the cotyledons and apical meristem through the soil. Once it reaches the surface, it straightens and pulls the cotyledons and shoot tip of the growing seedlings into the air. Beans, tamarind, and papaya are examples of plants that germinate this way.

Hypogeous

Another way of germination is hypogeous (or hypogeal), where the epicotyl elongates and forms the hook. In this type of germination, the cotyledons stay underground where they eventually decompose. Peas, for example, germinate this way.

Monocot Germination

In monocot seeds, the embryo's radicle and cotyledon are covered by a coleorhiza and coleoptile, respectively. The coleorhiza is the first part to grow out of the seed, followed by the radicle. The coleoptile is then pushed up through the ground until it reaches the surface. There, it stops elongating and the first leaves emerge.

Precocious Germination

While not a class of germination, precocious germination refers to seed germination before the fruit has released seed. The seeds of the green apple commonly germinate in this manner.

Pollen Germination

Another germination event during the life cycle of gymnosperms and flowering plants is the germination of a pollen grain after pollination. Like seeds, pollen grains are severely dehydrated before being released to facilitate their dispersal from one plant to another. They consist of a protective coat containing several cells (up to 8 in gymnosperms, 2-3 in flowering plants). One of these cells is a tube cell. Once the pollen grain lands on the stigma of a receptive flower (or a female cone in gymnosperms), it takes up water and germinates. Pollen germination is facilitated by hydration on the stigma, as well as by the structure and physiology of the stigma and style. Pollen can also be induced to germinate *in vitro* (in a petri dish or test tube). During germination, the tube cell elongates into a pollen tube. In the flower, the pollen tube then grows towards the ovule where it discharges the sperm produced in the pollen grain for fertilization. The germinated pollen grain with its two sperm cells is the mature male microgametophyte of these plants.

Self-Incompatibility

Since most plants carry both male and female reproductive organs in their flowers, there is a high risk of self-pollination and thus inbreeding. Some plants use the control of pollen germination as a way to prevent this self-pollination. Germination and growth of the pollen tube involve molecular signalling between stigma and pollen. In self-incompatibility in plants, the stigma of certain plants can molecularly recognise pollen from the same plant and prevent it from germinating.

Spore Germination

Germination can also refer to the emergence of cells from resting spores and the growth of sporeling hyphae or thalli from spores in fungi, algae and some plants. Conidia are asexual reproductive (reproduction without the fusing of gametes) spores of fungi which germinate under specific conditions. A variety of cells can be formed from the germinating conidia.

The most common are germ tubes which grow and develop into hyphae. Another type of cell is a conidial anastomosis tube (CAT); these differ from germ tubes in that they are thinner, shorter, lack branches, exhibit determinate growth and home toward each other. Each cell is of a tubular shape, but the conidial anastomosis tube forms a bridge that allows fusion between conidia.

Resting Spores

In resting spores, germination that involves cracking the thick cell wall of the dormant spore. For example, in zygomycetes the thick-walled zygosporangium cracks open and the zygospore inside gives rise to the emerging sporangiophore.

In slime molds, germination refers to the emergence of amoeboid cells from the hardened spore. After cracking the spore coat, further development involves cell division, but not necessarily the development of a multicellular organism (for example in the free-living amoebas of slime molds).

Ferns and Mosses

In plants such as bryophytes, ferns, and a few others, spores germinate into independent gametophytes. In the bryophytes (e.g., mosses and liverworts), spores germinate into protonemata, similar to fungal hyphae, from which the gametophyte grows. In ferns, the gametophytes are small, heart-shaped prothalli that can often be found underneath a spore-shedding adult plant.

Origin and Evolution

The origin of seed plants is a problem that still remains unsolved. However, more and more data tends to place this origin in the middle Devonian.

The description in 2004 of the proto-seed *Runcaria heinzelinii* in the Givetian of Belgium is an indication of that ancient origin of seed-plants. As with modern ferns, most land plants before this time reproduced by sending spores into the air, that would land and become whole new plants. The first "true" seeds are described from the upper Devonian, which is probably the theater of their true first evolutionary radiation. The seed plants progressively became one of the major elements of nearly all ecosystems.

Economic Importance

Edible Seeds

Further Information: List of Edible Seeds: Many seeds are edible and the majority of human calories comes from seeds, especially from cereals, legumes and nuts. Seeds also provide most cooking oils, many beverages and spices and some important food additives. In different seeds the seed embryo or the endosperm dominates and provides most of the nutrients. The storage proteins of the embryo and endosperm differ in their amino acid content and physical properties. For example the gluten of wheat, important in providing the elastic property to bread dough is strictly an endosperm protein.

Seeds are used to propagate many crops such as cereals, legumes, forest trees, turfgrasses and pasture grasses. Particularly in developing countries, a major constraint faced is the inadequacy of the marketing channels to get the seed to poor farmers. Thus the use of farmer-retained seed remains quite common.

Seeds are also eaten by animals, and are fed to livestock. Many seeds are used as birdseed.

Poison and Food Safety

While some seeds are edible, others are harmful, poisonous or deadly. Plants and seeds often contain chemical compounds to discourage herbivores and seed predators. In some cases, these compounds simply taste bad (such as in mustard), but other compounds are toxic or break down into toxic compounds within the digestive system. Children, being smaller than adults, are more susceptible to poisoning by plants and seeds.

A deadly poison, ricin, comes from seeds of the castor bean. Reported lethal doses are anywhere from two to eight seeds, though only a few deaths have been reported when castor beans have been ingested by animals.

In addition, seeds containing amygdalin—apple, apricot, bitter almond, peach, plum, cherry, quince, and others—when consumed in sufficient amounts, may cause cyanide poisoning. Other seeds that contain poisons include annona, cotton, custard apple, datura, uncooked durian, golden chain, horse-chestnut, larkspur, locoweed, lychee, nectarine, rambutan, rosary pea, sour sop, sugar apple, wisteria, and yew. The seeds of the strychnine tree are also poisonous, containing the poison strychnine. The seeds of many legumes, including the common bean (*Phaseolus vulgaris*), contain proteins called lectins which can cause gastric distress if the beans are eaten without cooking. The common bean and many others, including the soybean, also contain trypsin inhibitors which interfere with the action of the digestive enzyme trypsin. Normal cooking processes degrade lectins and trypsin inhibitors to harmless forms.

Other Uses

Cotton fibre grows attached to cotton plant seeds. Other seed fibres are from kapok and milkweed. Many important nonfood oils are extracted from seeds. Linseed oil is used in paints. Oil from jojoba and crambe are similar to whale oil. Seeds are the source of some medicines including castor oil, tea tree oil and the discredited cancer drug, Laetrile.

Many seeds have been used as beads in necklaces and rosaries including Job's tears, Chinaberry, rosary pea, and castor bean. However, the latter three are also poisonous.

Other seed uses include:

- Seeds once used as weights for balances.
- Seeds used as toys by children, such as for the game Conkers.
- Resin from *Clusia rosea* seeds used to caulk boats.
- Nematicide from milkweed seeds.
- Cottonseed meal used as animal feed and fertilizer.

Seed Records

Figure: *The massive fruit of the coco de mer.*

- The oldest viable carbon-14-dated seed that has grown into a plant was a Judean date palm seed about 2,000 years old, recovered from excavations at Herod the Great's palace on Masada in Israel. It was germinated in 2005. (A reported regeneration of *Silene stenophylla* (narrow-leafed campion) from material preserved for 31,800 years in the Siberian permafrost was achieved using fruit tissue, not seed.)

- The largest seed is produced by the coco de mer, or "double coconut palm", *Lodoicea maldivica.* The entire fruit may weigh up to 23 kilograms (50 pounds) and usually contains a single seed.
- The earliest fossil seeds are around 365 million years old from the Late Devonian of West Virginia. The seeds are preserved immature ovules of the plant *Elkinsia polymorpha.*

Recalcitrant Seed

Recalcitrant seeds (sometimes known as unorthodox seeds) are seeds that do not survive drying and freezing during ex-situ conservation. Moreover, these seeds cannot resist the effects of drying or temperatures less than 10° C; thus, they cannot be stored for long periods like *orthodox seeds* because they can lose their viability. Plants that produce recalcitrant seeds include avocado, mango, lychee, cocoa, rubber tree, some horticultural trees, and several plants used in traditional medicine such as species of Virola and Pentaclethra. Generally speaking, most tropical pioneer species have orthodox seeds but many climax species have recalcitrant or intermediate seeds.

Mechanisms of Damage

The two main mechanisms of action of damage to recalcitrant seeds are desiccation effect on the intracellular structures and the effect of metabolic damage from the formation of toxic chemicals such as free radicals. An example of the first type of damage would be found in some recalcitrant nontropical hardwood seeds, specifically the acorns of recalcitrant oaks, which can be stored in a nonfrozen state for up to two years provided that precautions be taken against drying. These seeds showed deterioration of cell membrane lipids and proteins after as few as 3-4 days of drying. Other seeds such as those of the sweet chestnut (*Castanea sativa*) show oxidative damage resulting from uncontrolled metabolism occurring during the drying process.

Seed Enhancement

Seed enhancement is a range of treatments of seeds that improves their performance after harvesting and conditioned, but before they are sown. They include priming, steeping, hardening, pregermination, pelleting, encrusting, film-coating, tagging and others, but excludes treatments for control of seed born pathogens. They are used to improve seed sowing, germination and seedling growth by altering seed vigor and/or the physiological state of the seed. The alteration may improve vigor or the physiological state of the seed by enhancing uniformity of germination. Treatments may include hydration treatments, such as priming, steeping, hardening, and pre-germination. Other treatment include the use of chemicals that trigger systemic acquired resistance or improve stress tolerance.

The use of antioxidants. Enhancements like pelleting, coating and encrusting improve seed handling and planting. Some treatments enhance nutrient availability or provide inoculates by delivering materials (other than pesticides) needed during sowing, germination and seedling establishment.

Seed Predation

Figure: *Strawberry fruit damaged by a mouse eating the seeds.*

Seed predation, often referred to as granivory, is a type of plant-animal interaction in which granivores (seed predators) feed on the seeds of plants as a main or exclusive food source, in many cases leaving the seeds damaged and not viable.

Granivores are found across many families of vertebrates (especially mammals, birds and insects) as well as invertebrates; thus, seed predation occurs in virtually all ecosystems. Seed predation is commonly divided into two distinctive temporal categories, pre-dispersal and post-dispersal predation, which may involve different strategies and requirements and have different implications at the individual and population level.

Nevertheless, regardless of timing, seed predation has great implications for plant population dynamics. To counterbalance effects of predation, plants have evolved defences such as seed morphology (size, shape, toughness) and chemical defences (secondary compounds such as tannins and alkaloids) to defend against their seed predators. However, as plants have adapted defences to seed predation (e.g., chemical compounds), so have seed predators adapted to plant defences (e.g., ability to detoxify chemical compounds). Thus, many interesting examples of coevolution arise from this dynamic relationship.

Seeds and Their Defences

Plant seeds are important sources of nutrition for animals across most ecosystems. Seeds contain food storage organs (e.g., endosperm) that provide nutrients to the developing plant embryo (cotyledon). This makes seeds an attractive food source for animals because they are a highly concentrated and localized nutrient source in relation to other plant parts.

Seeds of many plants have evolved a variety of defences to deter predation from seed predators. Plants may allocate resources toward physical or chemical defences. Seeds are often contained inside protective structures or fruit pulp that encapsulate seeds until they are ripe. Other physical defences include spines, hairs, fibrous seed coats and hard endosperm.

Some seeds have evolved strong anti-herbivore chemical compounds. In contrast to physical defences, chemical defences in seeds to deter consumption by seed predators by using chemicals that are toxic to granivores or inhibit the digestibility of the seed. These chemicals include non-protein amino acids, cyanogenic-gycosides, protease and amylase inhibitors, Phytohaemaglutinins. Seeds face trade-offs between allocation toward defences and the size and number of seeds produced. Additionally plants have trade-offs in seed defence allocation between deterring seed predators and attracting seed dispersers.

Plants may reduce the severity of seed predation by making seeds spatially or temporally scarce to granivores. Seed dispersal away from the parent plant is hypothesized to reduce the severity seed predation. Seed masting is an example of how plant populations are able to temporally regulate the severity of seed predation. Masting refers to a concerted abundance of seed production followed by a period of paucity. This strategy has the potential to regulate the size of the population of seed predators.

Pre- and Post-Dispersal Seed Predation

Seed predation often occurs at contrasting plant ontogenic stages. Predispersal seed predation takes place when seeds are removed from the parent plant before dispersal, and it has been most often reported in invertebrates, birds, and in granivorous rodents that clip fruits directly from trees and herbaceous plants. Postdispersal seed predation arises once seeds have been released from the parent plant. Birds, rodents, and ants are known to be among the most pervasive postdispersal seed predators. Furthermore, postdispersal seed predation can take place at two contrasting stages: predation on the "seed rain" and predation on the "seed bank". Whereas predation on the seed rain occurs when animals prey on released seeds usually flush with the ground surface, predation on the seed bank takes place after seeds have been incorporated deeply into the soil. Both pre- and post-dispersal seed predation are extremely frequent in nature. However, fundamental differences occur between the two. Pre-dispersal predators are most often specialists, as they are more likely to adapt to clustered resources. Predator specialisation requires that these animals utilise specific cues like plant chemistry (volatile compounds), colour and size to attack seeds. In addition, specialisation requires that the life cycle of the predator match the production of seeds by the host plant. Therefore, animals with shorter life span, like insects, are more common pre-dispersal predators. Common insect groups of pre-dispersal predators are Coleoptera, Hemiptera, Hymenoptera and Lepidoptera. For this reason, a greater diversity of predators feed on seeds prior to dispersal. Nevertheless, there are important vertebrate pre-dispersal predators, especially birds and small mammals.

Post-dispersal seed predation is extremely common in virtually all ecosystems. Given the heterogeneity in both resource type (seeds from different species are available), quality (seeds of different ages and/or different status of integrity or decomposition) and location (seeds are scattered and hidden in the environment), most post-dispersal predators have generalist habits. These predators usually belong to a diverse array of animals, such as ants, beetles, crabs, fish, rodents and birds. In addition, not only is the assemblage of post-dispersal seed predators diverse, but its composition varies considerably among ecosystems.

The severity of predation rates also varies between pre- and post-dispersal predation. Although both types of predation can be extremely high, studies simultaneously evaluating both pre- and post-dispersal seed predation suggest that the latter is often more severe than the former. However, seeds already dispersed and not consumed have a chance to germinate, and seeds consumed could die regardless of predation due to constraints imposed by unsuitable germination sites. On the other hand, all non-dispersed seeds predated would be no more than a waste of energy by the parent plant. Therefore, the demographic impacts of pre-dispersal predation on plant populations are likely to be stronger than the impacts of post-dispersal predation.

Effects on Plant Demography

The complex relationship between seed predation and plant demography is an important topic of plant-animal interactive studies. Plant population structure and size over time is closely associated with the effectiveness at which seed predators locate, consume, and disperse seeds. In many cases this relationship is dependent upon the type of seed predator (specialist vs. generalist) or the particular habitat in which the interaction is taking place. The role of seed predation on plant demography may be either detrimental or in particular cases actually beneficial to plant populations.

The Janzen-Connell model is one of the most important studies of how seed predation may influence plant demography. The Janzen-Connell model illustrates how seed density and survival respond to distance from the parent tree and differential rates of seed predation. Seed density is hypothesized to decrease as distance from the parent tree increases. Where seeds are most abundant under the parent tree, seed predation is predicted to be at its highest. As distance from the parent tree increases, seed abundance and thus seed predation are predicted to decrease as seed survival increases.

The degree to which seed predation influences plant populations may vary by whether a plant species is safe site limited or seed limited. If a population is safe site limited it is likely that seed predation will have little impact to the success of the population. In safe site limited populations increased seed abundance does not translate into increased seedling recruitment. However, if a population is seed limited, seed predation has a better chance of negatively affecting the plant population by decreasing seedling recruitment. A study by Maron and Simms found both safe site limited and seed limited populations depending on the habitat in which the seed predation was taking place. In dune habitats seed predators (Deer mice) were limiting seedling recruitment in the population, thus negatively affecting the population. However, in grassland habitat the seed predator had little effect on the plant population because it was safe site limited.

In many cases seed predators support plant populations by dispersing seeds away from the parent plant. By dispersing seeds away from the parent plant the seed predator is in effect supporting gene flow between populations. Other seed predators collect seeds and then store or cache them for later consumption. In the case that the seed predator is unable to locate the buried or hidden seed there is a chance that it will later germinate and grow, supporting the species dispersal.

Seed Testing

Seed testing is performed in dedicated laboratories by trained and usually certified analysts. The tests are designed to evaluate the quality of the seed lot being sold. Several tests are done:

- Germination test: Reports the percentage of seed that germinated. Tests are usually made in 200 or 400 seed samples.
- Purity test: The percentage of seed described on the label that is actually found in the quantity of seed.

- TZ test: A test for viability that involves three steps: 1. preconditioning (imbibition); 2. preparation and staining (sometimes cutting the seed and then soaking the seed in a 2,3,5 triphenyl tetrazolium chloride solution); and 3. evaluation (examining the seed for a colour change in the embryo).
- Weed test: Examines a sample of seed and identifies every seed that is different from the labelled seed kind.

Seed Trap

Seed traps are used in ecology and forestry to capture seeds falling from plants, allowing seed production and dispersal to be quantified. They come in several forms, including funnel traps, sticky traps (using materials such as fly paper), nets and pots exposed in the field.

Types of Trap

There are many options when using seed traps based on the specific need for the project. Seed traps can be made in different sizes, shapes, and of different material. Traditionally, seed traps are wooden frames with a screened bottom. Traps with metal frames have also been used. Additionally, funnel-shaped traps that are set above the soil or leveled at the ground, traps with screen or cloth bags, traps with water or soil to germinate them, plastic buckets, or traps with sticky surfaces have been used.

Sticky Traps

The substance on sticky traps must be non toxic to the seeds and non-drying. Many sticky traps are petroleum-based. They are also cheaper (approximately $1.20-1.30 per use), lighter and less bulky than other traps. However, rain or dew may affect the adhesive. Heat and light intensities can also cause sticky traps to loos their adhesiveness.

Pollen Traps

Pollen traps are used to measure production and dispersal of pollen in plants. These traps are commonly made of glass slides with silicon oils or sticky tapes. They can be set up horizontally or as cylinder containers.

Fruit Traps

These traps can collect fruit production from various plants. These are also constructed on the ground, hanging or as basins or funnels.

Choosing a Trap

When choosing a trap, certain factors must be considered. The seed dispersal unit (where the seeds will fall), timing of dispersal, and density of seed fall (how much will be produced). Smaller traps may be appropriate for trees that produce more seeds, and larger traps may be appropriate for trees that produce less seeds to guarantee collection. It is difficult to compare different traps. There are also some drawbacks to consider when choosing a trap. Wood traps may easily deteriorate if not constructed properly. Traps may also be hard to store or set up depending on size. Cloth and screens are easily torn in bad weather conditions and are also targets for animals. Screens may also harm many insects. Traps with water or soil must be maintained consistently. Extra seeds may clog the traps or seeds may be blown or washed out causing an over or underestimate of seed despersal. Sticky traps are also clogged by other debris.

Maintenance

Traps must be checked weekly depending on the rate of seed despersal. Seed identification must be accurate. This can be challenging because there is no comprehensive identification key, so focusing on only a few species at a time is important. You can also attempt to germinate the seeds. Screens may help preditation from animals and insects.

Seed Legislation

Introduction

The basic purpose of seed legislation and its subsequent enforcement is to regulate the quality of seed sold to farmers.

Types

Seed legislation can broadly be divided into two groups:

1. Sanctioning legislation
2. Control legislation

Seed Legislation in India

The rapid development of agricultural production with the introduction of hybrid varieties of maize, jowar and bajra, and dwarf varieties of wheat and paddy, however, necessitated the enactment of seed legislation. On 29th December 1966, the Seeds Act was passed. It came into force throughout the country on 2nd October 1969.

The main features of the Seeds Act, 1966

1. *Applicability:* It is applicable only to notified kinds/varieties of seed and vegetatively propagating materials used for sowing.
2. *Sanctioning legislation:* The Act provides for the formation of an apex advisory body, namely, the Central Seed Committee; the Central Seed Certification Board; establishment of Seed Certification Agencies; and Central and State Seed Testing Laboratories, etc.

3. *Regulatory legislation:* The Act provides for the provisions for notification of kinds/varieties to be brought under the purview of the Seeds Act; regulation regarding the sale of seed; and the establishment of a suitable seed law enforcement machinery. Under the Act the Central Govt. is empowered to make rules to carry out the purposes of the Act and to give directions to State Govt. It necessary, for carrying into execution, in the state concerned the provisions of the Act or Rules.

Regulatory Legislations

Notification of Kinds or Varieties

The Central Government after consultation with the Central Seed Committee can issue notification for such kinds/varieties for which it is of the opinion that it has become necessary or expedient to regulate the quality of seed to be sold for the purposes of agriculture. The notification brings such kinds/varieties under the purview of the Seeds Act.

Regulation of Sale of Notified Kinds/Varieties

Under Section 7 of the Act, the seeds of notified kinds/varieties can be sold in containers only, if they are either (a) labelled or, (b) certified and meet the minimum prescribed requirements.

Statutory Requirements for Sale of Seed

Labelled Seed

The seed container shall be labelled in the following manner. The colour of the label shall be buff.

1. Kind
2. Variety
3. Lot number
4. Date of test
5. Inert matter percentage
6. Pure seed percentage
7. Other crop seed percentage
8. Weed seeds percentage
9. Germination percentage
10. Net content
11. Seller's name and address
12. If treated, then either of the following two statements should appear on the label.

Other Requirements

1. No person can sell, keep for sale, and offer to sell, barter or otherwise supply any seed of any notified kind/variety after the validity period recorded on the container, mark or label has expired.
2. No person can alter obliterate or deface any mark or label attached to the container of any seed.
3. Every person selling, keeping for sale, offering to sell, bartering or otherwise supplying any seed of notified kind or variety under Section 7 shall keep over a period of three years a complete record of each lot of seed sold, except for this, any seed sample may be discarded one year after the entire lot represent by such sample has been disposed of.

Kinds or Varieties

Under Section 17 of the Act, no person shall, for the purpose of sowing or planting by any person, export or import or cause to be exported or imported any seed of any notified kind or variety, unless:

a. It conforms to the minimum limits of germination and purity specified for that seed under clauses (a) of Section 6; and
b. Its container bears, in the prescribed manner, the mark or label with the correct particulars there have specified for that seed under clauses (b) of Section 6.

Sale of Seed

The seed inspectors, possessing prescribed qualification, appointed by the State Governments through notification in the official Gazette are responsible for Seed Law Enforcement in their respective areas. The State Governments will also notify State Seed Laboratories and State Analysts where the samples drawn by seed in drawn by seed inspectors can be tested under the supervision of the State Seed Analyst.

Penalties for Offenders

The person who contravenes any provision of the Seeds Act or any rule thereunder, or prevents a seed inspector from taking a sample under this Act, or prevents a seed inspector from excising any other power conferred on him by, or under, the Seeds Act, on conviction is punishable:

1. For the first offence with a fine which may extend to five hundred rupees; and
2. In the event of such a person having been previously convicted of an offence under this section, with imprisonment for a term which may extend upto six months, or with a fine which may extend to one thousand rupees, or with both.

3. Forfeiting of property. When any person has been convicted under this Act, for the contravention of any of the provisions of this Act or the Rules made thereunder, the seed in respect of which the contravention has been committed may be forfeited to the Government.
4. Offences by companies. Where an offence under this Act has been committed by a company every person who, at the time the offence was committed, was in charge of, and was responsible to the company for the conduct of the business of the company as well as the company, shall be deemed to be guilty of the offence and shall be liable to be proceeded against and punished accordingly.

Protect Your Brain and Safeguard Your Ideas

Gone are the days when people used to believe in phrases like "knowledge is the best of all wealth, it is easy to carry, thieves cannot steal it, the tyrants cannot seize it, and neither water nor fire can destroy it"-Naladiyar.

In the present global scenario, before even you protect your personal account number, you have to ensure about the confidentiality of your computer's password. The public as well as private organizations are investing huge amount of money for maintaining the security of their information.

The world is revolving around the novel ideas, innovations, technologies and ultimately money. The moment you put down ideas on the paper or a machine, in simpler way, in a tangible form, "BEWARE", it needs to be protected from being copied.

Most treasured asset then, now and always is "IDEAS, CREATIONS, INNOVATIONS" which collectively form "intellectual property". IP law gives exclusive rights on the creation to the creator for certain period of time as a token of reward for the inventor and also as an incentive for others.

The innovation/creation enters into public domain after the termination of the specified time period under IP law. All of our life is surrounded by IP in one or the other way.

It covers a wide range of products from small screw to big machine, from genes to microorganisms, from life saving drugs to cosmetics, mere name of an organization or trade union to a logo of the company, a feature film, documentary, music, songs and the list goes on.

There are numerous acts and laws in different countries that offer protection under various situations. The Director General of WIPO, Francis Gurry in his opening statement in the Assemblies of the Member States of WIPO held during September 22 to October 1, 2009 reports that 1.85 million patent applications, 3.3 million trademark applications and 621,000 industrial design applications were filed around the world in the year 2007.

India and IPR

Though Indian Patent laws were first promulgated in 1856 and new patent law was introduced after independence in 1970, India was not keen on enforcement of Intellectual Property Rights until recently. Being a member of World Trade Organization (WTO), it was compelled to implement WTO-standard IPR protection laws by 2005 and was given a time span of ten years from 1995 to 2005 to embrace the necessary modifications to its existing IP regime. The US has placed India in priority watch list for not providing sufficient level of IP protection. IPR in India comprises patents, copyright, designs, trademark, layout designs of integrated circuits, geographical indication and plant variety protection. There are separate legislations for each of these rights that specifies the rules and regulations, criteria for registration and stipulates the time period. Despite the fact that stronger IPRs encourage innovations, India is yet to enforce an effective IP setting. As the ruler so the ruled. The Indian companies are lagging in the number of patents filed. According to the annual report-2007/08 35,218 patents, 1,23,514 trademarks and 6,402 designs have been filed in the Indian Patent Office of which 15,261 patents, 1,00857 trademarks and 4,928 designs were granted. The number of patent applications when compared to other two Asian countries, China and Japan is awfully low. Thompson Reuters Patent Focus report-2009 states that the State Intellectual Property Office (SIPO) in China, received 613,922 patents from Chinese entities between January to November, 2008 and 396,291 patent applications were received by Japan Patent Office (JPO).

Forms of IPR

Patent is an exclusive right to use, sell or to license the invention, granted by the country to the inventor for maximum of 20 years from the date of filing in exchange for a public disclosure of the invention. It includes inventions in all fields of human endeavour and scientific discoveries like drugs, machines, automobiles, transgenics. The most important requirement is functionality of the invention apart from novelty. The recent patent litigation in India include the rejection of the patent filed by Novartis on anticancer drug– "Gleevac" and TVS v/s Bajaj on Digital Twin Spark Plug Ignition [DTS-i] technology. The Department of Science and Technology under Technology Information Forecasting and Assessment Council (TIFAC) has set up Patent Facilitating Centre in 1995. PFC contributes first Indian patent searchable databases EKASWA-A, EKASWA-B, EKASWA-C which provides information on patent applications filed, published and notified for opposition. Copyrights are group of rights that include publication, distribution and adaptation granted in the field of literary, artistic and scientific works like stories, dialogues, books, articles, paintings, cartoons, pictures, movies, songs, music and also software programmes. Related rights covers performance of performing artists, phonograms and broadcasts over television, radio, and stage performances etc., Copyright is granted upon mere creation and the rights can be exercised over a specified time period in relation to the type of work, after which it enters public domain. Government of Karnataka has taken measures to combat video and audio cassette piracy by introducing Goonda Act.

Industrial design rights are the rights that protect the visual design and appearance of the novel creations of objects like shape, colour and configuration. Unlike patents, these rights are confined to outer look of an object and are valid for a period of 10 years from the date of filing. Examples for industrial designs include mobile handsets, automobile designs, fabrics, etc.

Trademark and trade name are the most well-known form of IPR. It comprises symbols, letters, words, logos, pictures, sound, combination of colours or any other such indications which are aimed to distinguish between different products or manufacturers.

Main purpose of the trademark is to avoid confusion and deception among buyers. The rights offered under these IP are for unlimited time period provided it is renewed for every ten years. For example Bata, Paragon, Boots are the trademark of different manufacturers of footwear.

Geographical Indications are the rights enjoyed by a group of people for a good originating from a geographical area possessing the distinct features attributable to the geographic area of origin. This right can also be renewed for every 10 years but it cannot be licensed. Famous GI include Kolhapuri chappals, Nagpur Oranges, Agra Petha, Assam Tea, Kanchipuram Sarees, etc., There are 61 registered GI's in India of which 21 are from Karnataka including Mysore Silks, Mysore Rosewood inlay, Channapatna Toys, Nanjangud Banana etc. Most recent addition to this list is Tirupathi laddu which is facing criticisms for being recognised as GI. Rights of Plant breeders and farmers are shielded under Plant Varieties and Farmers' Rights Protection which includes development of new and distinct varieties. Semiconductor Integrated Circuit Layout Designs Act protects layout designs of integrated circuits and is binding for a period of 10 years THE SEEDS ACT, 1966 (ACT NO. 54 OF 1966) [29th December, 1966] An Act to provide for regulating the quality of certain seeds for sale, and for matters connected therewith BE it enacted by Parliament in the Seventeenth Year of the Republic of India as follows:

Short title, extent and commencement 1: (1) This Act may be called the Seeds Act, 1966. (2) It extends to the whole of India. (3) It shall come into force on such date as the Central Government may, by notification in the Official Gazette, appoint, and different dates may be appointed for different provisions of this Act, and for different States or for different areas thereof.

Definitions 2: In this Act, unless the context otherwise requires, (1) "Agriculture" includes horticulture; (2) "Central Seed Laboratory" means the Central Seed Laboratory established or declared as such under sub-section (1) of section 4; (3) "Certification agency" means the certification agency established (4) "Committee" means the Central Seed Committee constituted under sub-section (1) of Section 3; (5) "Container" means a box, bottle, casket, tin, barrel, case, receptacle, sack, bag, wrapper or other thing in which any article or thing is placed or packed; (6) "Export" means taking out of India to a place outside India; (7) "Import" means bringing into India from a place outside India; (8) "Kind" means one or more related species or sub-species of crop plants each individually or collectively known by one common name such as cabbage, maize, paddy and wheat; (9) "notified kind or variety" , in relation to any seed, means any kind or variety thereof notified under Section 5; (10) "Prescribed" means prescribed by rules made under this act. (11) "seed" means any of the following classes of seeds used for sowing or planting : (i) seeds of food crops including edible oil seeds and seeds of fruits and vegetables; (ii) cotton seeds; (iii) seeds of cattle fodder; and includes seedlings, and tubers, bulbs, rhizomes, roots, cuttings, all types of grafts and other vegetatively propagated material, of food crops or cattle fodder; (12) "Seed Analyst" means a Seed Analyst appointed under section 12; (13) "Seed Inspector" means a Seed Inspector appointed under section 13; (14) "State Government", in relation to a Union territory, means the administrator thereof; (15) "State Seed Laboratory", in relation to any State, means the State Seed Laboratory established or declared as such under sub-section (2) of section 4 for that State; and (16) "variety" means a sub-division of a kind identifiable by growth, yield, plant, fruit, seed, or other characteristic.

Central Seed Committee 3: (1) The Central Government shall, as soon as may be after the commencement of this Act, constitute a Committee called the Central Seed Committee to advise the Central Government and the State Governments on matters arising out of the administration of this Act and to carry out the other functions assigned to it by or under this Act. (2) The Committee shall consist of the following members, namely:- (i) a Chairman to be nominated by the Central Government; (ii) eight persons to be nominated by the Central Government to represent such interests that Government thinks fit, of whom not less than two persons shall be representatives of growers of seed; (iii) one person to be nominated by the Government of each of the States. (3) The members of the Committee shall, unless their seats become vacant earlier by resignation, death or otherwise, be entitled to hold office for two years and shall be eligible for renomination. (4) The Committee may, subject to the previous approval of the Central Government, make bye-laws fixing the quorum and regulating its own procedure and the conduct of all business to be transacted by it. (5) The Committee may appoint one or more sub-committees, consisting wholly of members of the Committee or wholly of other persons or partly of members of the Committee and partly of other persons, as it thinks fit, for the purpose of discharging such of its functions as may be delegated to such sub-committee or sub-committees by the Committee. (6) The functions of the Committee or any sub-committee thereof may be exercised notwithstanding any vacancy therein. (7) The Central Government shall appoint a person to be the secretary of the Committee and shall provide the Committee with such clerical and other staff as the Central Government considers necessary.

Central Seed Laboratory and State Seed Laboratory 4: (1) The Central Government may, by notification in the Official Gazette, establish a Central Seed Laboratory or declare any seed laboratory as the Central Seed Laboratory to carry out the functions entrusted to the Central Seed Laboratory by or under this Act. (2) The State Government may, by notification in the Official Gazette, establish one or more State Seed Laboratories or declare any seed laboratory as a State Seed Laboratory where analysis of seeds of any notified kind or variety shall be carried out by Seed Analysts under this Act in the prescribed manner.

Power to notify kinds or varieties of seeds 5: If the Central Government, after consultation with the committee, is of opinion that it is necessary or expedient to regulate the quality of seed of any kind or variety to be sold for purposes of agriculture, it may, by notification in the Official Gazette, declare such kind or variety to be a notified kind or variety for the purposes of the Act and different kinds or varieties may be notified for different States or for different areas thereof.

Power to specify minimum limits of germination and purity, etc. 6: The Central Government may, after consultation of the Committee and by notification in the Official Gazette, specify – (a) the minimum limits of germination and purity with respect to any seed of any notified kind or variety: (b) the mark or label to indicate that such seed conforms to the minimum limits of germination and purity specified under clause (a) and the particulars which marks or label may contain.

Regulation of sale of seeds of notified kinds or varieties 7: No person shall, himself or by any other person on his behalf, carry on the business of selling, keeping for sale, offering to sell, bartering or otherwise supplying any seed of any notified kind or variety, unless- (a) such seed is identifiable as to its kind or variety; (b) such seed conforms to the minimum limits of germination and purity specified under clause (a) of section 6; (c) the container of such seed bears in the prescribed manner, the mark or label containing the correct particulars thereof, specified under clause (b) of section 6; and (d) he complies with such other requirements as may be prescribed.

Certification agency 8: The State Government or the Central Government in consultation with the State Government may, by notification in the Official Gazette, establish a certification agency for the State to carry out the functions entrusted to the certification agency by or under this Act.

Grant of certificate by certification agency 9: (1) Any person selling, keeping for sale, offering to sell, bartering or otherwise supplying any seed of any notified kind or variety may, if he desires to have such seed certified by the certification agency, apply to the certification agency for the grant of a certificate for the purpose. (2) Every application under sub-section (1) shall be made in such form, shall contain such particulars and shall be accompanied by such fees as may be prescribed. (3) On receipt of any such application for the grant of a certificate, the certification agency may, after such enquiry as it thinks fit and after satisfying itself that the seed to which the application relates conforms to the minimum limits of germination and purity specified for that seed under clause (a) of section 6, grant a certificate in such form and on such conditions as may be prescribed.

Revocation of certificate 10: If the certification agency is satisfied, either on a reference made to it in this behalf or otherwise, that- (a) the certificate granted by it under section 9 has been obtained by misrepresentation as to an essential fact; or (b) the holder of the certificate has, without reasonable cause, failed to comply with the conditions subject to which the certificate has been granted or has contravened any of the provisions of this Act or the rules made thereunder; then, without prejudice to any other penalty to which the holder of the certificate may be liable under this Act, the certification agency may, after giving the holder of the certificate an opportunity of showing cause, revoke the certificate.

Appeal 11: (1) Any person aggrieved by a decision of a certification agency under section 9 or section 10, may, within thirty days from the date on which the decision is communicated to him and on payment of such fees as may be prescribed, prefer an appeal to such authority as may be specified by the State Government in this behalf: Provided that the appellate authority may entertain an appeal after the expiry of the said period of thirty days if it is satisfied that the appellate was prevented by sufficient cause from filing the appeal in time. (2) On receipt of an appeal under sub-section (1), the appellate authority shall, after giving the appellant an opportunity of being heard, dispose of the appeal as expeditiously as possible. (3) Every order of the appellate authority under this section shall be final.

Seed Analysts 12: The State Government may, by notification in the Official Gazette, appoint such persons as it thinks fit, having the prescribed qualifications, to be Seed Analysts and define the areas within which they shall exercise jurisdiction.

Seed Inspectors 13: (1) The State Government may, by notification in the Official Gazette, appoint such persons as it thinks fit, having the prescribed qualifications, to be Seed Inspectors and define the areas within which they shall exercise jurisdiction. (2) Every Seed Inspector shall be deemed to be a public servant within the meaning of section 21 of the Indian Penal Code (45 of 1860) and shall be officially subordinate to such authority as the State Government may specify in this behalf.

Powers of Seed Inspector 14: (1) The Seed Inspector may- (a) take samples of any seed of any notified kind or variety from- (i) any person selling such seed; or (ii) any person who is in the course of conveying, delivering or preparing to deliver such seed to a purchaser or a consignee; or (iii) a purchaser or a consignee after delivery of such seed to him; (b) send such sample for analysis to the Seed Analyst for the area within which such sample has been taken; (c) enter and search at all reasonable times, with such assistance, if any, As he considers necessary, any place in which he has reason to believe that an offence under this Act has been or is being committed and order in writing the person in possession of any seed in respect of which the offence has been or is being committed, not to dispose of any stock of such seed for a specific period not exceeding thirty days or, unless the alleged offence is such that the defect may be removed by the possessor of the seed, seize the stock of such seed; (d) examine any record, register, document or any other material object found in any place mentioned in clause (c) and seize the same if he has reason to believe that it may furnish evidence of the commission of an offence punishable under this Act; and (e) exercise such other powers as may be necessary for carrying out the purposes of this Act or any rule made thereunder. (2) Where any sample of any seed of any notified kind or variety is taken under clause (a) of sub-section (1), its cost, calculated at the rate at which such seed is usually sold to the public, shall be paid on demand to the person from whom it is taken. (3) The power conferred by this section includes power to break-open any container in which any seed of any notified kind or variety may be contained or to break-open the door of any premises where any such seed may be kept for sale: Provided that the power to break-open the door shall be exercised only after the owner or any other person in occupation of the premises, if he is present therein, refuses to open the door on being called upon to do so. (4) Where the Seed Inspector takes any action under clause (a) of sub-section (1), he shall, as far as possible, call not less than two persons to be present at the time when such action is taken and take their signatures on a memorandum to be prepared in the prescribed form and manner. (5) The provisions of the Code of Criminal Procedure, 1898 (5 of 1898), shall, so far as may be, apply to any search or seizure under this section as they apply to any search or seizure made under the authority of a warrant issued under section 98 of the said Code.

Procedure to be followed by Seed Inspectors 15: (1) Whenever a Seed Inspector intends to take sample of any seed of any notified kind or variety for analysis, he shall- (a) give notice in writing, then and there, of such intention to the person from whom he intends to take sample; (b) except in special cases provided by rules made under this Act, take three representative samples in the prescribed manner and mark and seal or fasten up each sample in such manner as its nature permits. (2) When samples of any seed of any notified kind or variety are taken under sub-section (1), the Seed Inspector shall- (a) deliver one sample to the person from whom it has been taken; (b) send in the prescribed manner another sample for analysis to the Seed Analyst for the area within which such sample has been taken; and (c) retain the remaining sample in the prescribed manner for production in case any legal proceedings are taken or for analysis by the Central Seed Laboratory under sub-section (2) of section 16, as the case may be. (3) If the person from whom the samples have been taken refuses to accept one of the samples, the Seed Inspector shall send intimation to the Seed Analyst of such refusal and thereupon the Seed Analyst receiving the sample for analysis shall divide it into two parts and shall seal or fasten up one of those parts and shall cause it, either upon receipt of the sample or when he delivers his report, to be delivered to the Seed Inspector who shall retain it for production in case legal proceedings are taken. (4) Where a Seed Inspector takes any action under clause (c) of sub-section (1) of section 14: (a) he shall use all despatch in ascertaining whether or not the seed contravenes any of the provisions of section 7 and if it is ascertained that the seed does not so contravene, forthwith revoke the order passed under the said clause or, as the case may be, take such action as may be necessary for the return of the stock of the seed seized; (b) if he seizes the stock of the seed, he shall, as soon as may be, inform a magistrate and take his orders as to the custody thereof; (c) without prejudice to the institution of any prosecution, if the alleged offence is such that the defect may be removed by the possessor of the seed, he shall, on being satisfied that the defect has been so removed, forthwith revoke the order passed under the said clause. (5) Where as Seed Inspector seizes any record, register, document or any other material object under clause (d) of sub-section (1) of section 14, he shall, as soon as may be, inform a magistrate and take his orders as to the custody thereof.

Report of Seed Analyst 16: (1) The Seed Analyst shall, as soon as may be after the receipt of the sample under sub-section (2) of section 15, analyse the sample at the State Seed Laboratory and deliver, in such form as may be prescribed, one copy of the report of the result of the analysis to the Seed Inspector and another copy thereof to the person from whom the sample has been taken. (2) After the institution of a prosecution under this Act, the accused vendor or the complainant may, on payment of the prescribed fee, make an application to the court for sending any of the samples mentioned in clause (a) or clause (c) of sub-section (2) of section 15 to the Central Seed Laboratory for its report and on receipt of the application, the court shall first ascertain that the mark and the seal or fastening as provided in clause (b) of sub-section (1) of section 15 are intact and may then despatch the sample under its own seal to the Central Seed Laboratory which shall thereupon send its report to the court in the prescribed form within one month from the date of receipt of the sample, specifying the result of the analysis. (3) The report sent by the Central Seed Laboratory under sub-section (2) shall supersede the report given by the Seed Analyst under sub-section (1). (4) Where the report sent by the Central Seed Laboratory under sub-section (2) is produced in any proceedings under Section 19, it shall not be necessary in such proceedings to produce any sample or part thereof taken for analysis.

Restriction on export and import of seeds of notified kinds or varieties 17: No person shall, for the purpose of sowing or planting by any person (including himself), export or import or cause to be exported or imported any seed of any notified kind or variety, unless- (a) it conforms to the minimum limits of germination and purity specified for that seed under clause (a) of section 6; and (b) its container bears, in the prescribed manner, the mark or label with the correct particulars thereof specified for that seed under clause (b) of section 6.

Recognition of seed certification agencies of foreign countries 18: The Central Govt. may, on the recommendation of the Committee and by notification in the Official Gazette, recognise any seed certification agency established in any foreign country, for the purposes of this Act.

Penalty 19: If any person- (a) contravenes any provision of this Act or any rule made thereunder; or (b) prevents a Seed Inspector from taking sample under this Act; or (c) prevents a Seed Inspector from exercising any other power conferred on him by or under this Act; he shall, on conviction, be punishable- (i) for the first offence with fine which may extend to five hundred rupees, and (ii) in the event of such person having been previously convicted of an offence under this section, with imprisonment for a term which may extend to six months, or with fine which may extend to one thousand rupees, or with both.

Forfeiture of Property 20: When any person has been convicted under this Act for the contravention of any of the provisions of this Act or the rules made thereunder, the seed in respect of which the contravention has been committed may be forfeited to the Government.

Offences by Companies 21: (1) Where an offence under this Act has been committed by a company, every person who at the time the offence was committed was in charge of, and was responsible to the company for the conduct of the business of the company, as well as the company, shall be deemed to be guilty of the offence and shall be liable to be proceeded against and punished accordingly: Provided that nothing contained in this sub-section shall render any such person liable to any punishment under this Act if he proves that the offence was committed without his knowledge and that he exercised all due diligence to prevent the commission of such offence. (2) Notwithstanding anything contained in sub-section (1), where an offence under this Act has been committed by a company and it is proved that the offence has been committed with the consent or connivance of, or is attributable to any neglect on the part of, any director, manager, secretary or other officer of the company, such director, manager, secretary or other officer shall also be deemed to be guilty of that offence and shall be liable to be proceeded against and punished accordingly. Explanation.– For the purpose of this section,- (a) "company" means any body corporate and includes a firm or other association of individuals; and (b) "director", in relation to a firm, means a partner in the firm.

Protection of action taken in good faith 22: No suit, prosecution or other legal proceeding shall lie against the Government or any officer of the Government for anything which is in good faith done or intended to be done under this Act.

Power to give directions 23: The Central Government may give such directions to any State Government as may appear to the Central Government to be necessary for carrying into execution in the State any of the provisions of this Act or of any rule made thereunder.

Exemption 24: Nothing in this Act shall apply to any seed of any notified kind or variety grown by a person and sold or delivered by him on his own premises direct to another person for being used by that person for the purpose of sowing or planting.

Power to make rules 25: (1) The Central Government may, by notification in the Official Gazette, make rules to carry out the purpose of this Act. (2) In particular and without prejudice to the generality of the fore-going power, such rules may provide, for- (a) the functions of the Committee and the travelling and daily allowances payable to members of the Committee and members of any subcommittee appointed under sub-section (5) of section 3; (b) the functions of the Central Seed Laboratory; (c) the functions of a certification agency; (d) the manner of marking or labelling the container of seed of any notified kind or variety under clause (c) of Section 7 and under clause (b) of section 17; (e) the requirements which may be complied with by a person carrying on the business referred to in section 7; (f) the form of application for the grant of a certificate under section 9, the particulars it may contain, the fees which should accompany it, the form of the certificate and the conditions subject to which the certificate may be granted; (g) the form and manner in which and the fee on payment of which an appeal may be preferred under section 11 and the procedure to be followed by the appellate authority in disposing of the appeal; (h) the qualifications and duties of Seed Analysts and Seed Inspectors; (i) the manner in which samples may be taken by the Seed Inspector, the procedure for sending such samples to the Seed Analyst or the Central Seed Laboratory and the manner of analysing such samples; (j) the form of report of the result of the analysis under sub-section (1) or sub-section (2) of section 16 and the fees payable in respect of such report under the said sub-section (2); (k) the records to be maintained by a person carrying on the business referred to in section 7 and the particulars which such records shall contain; and (l) any other matter which is to be or may be prescribed. (3) Every rule made under this Act shall be laid as soon as may be after it is made, before each House of Parliament while it is in session for a total period of thirty days which may be comprised in one session or in two successive sessions, and if, before the expiry of the session in which it is so laid or the session immediately following, both Houses agree in making any modification in the rule or both Houses agree that the rule should not be made, that rule shall, thereafter have effect only in such modified form or be of no effect, as the case may be; so however, that any such modification or annulment shall be without prejudice to the validity of anything previously done under that rule. Composition of the Committee. 5. Powers and functions of the Committee. 6. Powers of Committee to specify minimum limits of germination, purity, seed health, etc. 7. Registration and other Sub-Committees of the Committee and their functions. 8. Procedure of the Committee and its Sub-Committees. 9. Secretary and other officers of the Committee. 13. Registration of seeds of any kind or varieties. 14. Procedure for registration. 15. Special provision for registration of transgenic varieties. 16. Cancellation of registration of seeds of kinds and varieties. 17. Notification of cancellation of registration of seeds of kinds and varieties. 18. Exclusion of certain kinds or varieties of seed from registration. 19. Evaluation of performance. 20. Compensation to farmers. 21. Seed producers and processing units to be registered. 22. Seed dealers to be registered. 23. Horticulture nursery to be registered. 24. Duties of registration holder of horticulture nursery.

Chapter IV Regulation Of Sale Of Seed And Seed Certification Agencies: 25. Regulation of sale of seeds of registered kinds and varieties. 26. State Seed Certification Agency. 27. Accreditation of Seed Certification Agencies. 28. Grant of certificate by the State Seed Certification Agency. 29. Revocation of certificate. 30. Recognition of seed certification agencies in foreign countries.

Chapter V Appeals: 31. Appeals.

Chapter VI Seed Analysis And Seed Testing: 32. Central and State Seed Testing Laboratories. 33. Seed Analysts. 34. Seed Inspectors. 35. Powers of Seed Inspector.

Chapter VII Export And Import Of Seeds: 36. Import of seeds. 37. Export of seeds.

Chapter VIII Offences And Punishment: 38. Offences and punishment. 39. Forfeiture of property. 40. Offences by companies.

Chapter IX Power Of Central Government: 41. Power of the Central Government to give directions to the State Governments. 42. Power of the Central Government to issue directions to the Committee. 43. Exemptions from registration

Chapter X Miscellaneous: 44. Protection of action taken in good faith 45. Power to remove difficulties. 46. Power of Central Government to make rules. 47. Power of the Committee to make regulations. 48. Rules and regulations to be laid before Parliament. 49. Repeal and savings. THE SCHEDULE THE SEEDS BILL, 2004 A BILL to provide for regulating the quality of seeds for sale, import and export and to facilitate production and supply of seeds of quality and for matters connected therewith or incidental thereto. BE it enacted by Parliament in the Fifty-Fifth Year of the Republic of India as follows:

Chapter I Preliminary: Short title, extent, application and commencement.

1. (1) This Act may be called the Seeds Act, 2004. (2) It extends to the whole of India. (3) Save as otherwise provided in this Act, it shall apply to- (a) every dealer; and (b) every producer of seed except when the seed is produced by him for his own use and not for sale. (4) It shall come into force on such date as the Central Government may, by notification, appoint. Provided that different dates may be appointed for different provisions of this Act and any reference in any such provision to the commencement of this Act shall be construed as a reference to the coming into force of that provision. Definitions.
2. In this Act, unless the context otherwise requires, - (1) "agriculture" includes horticulture, forestry and cultivation of plantation, medicinal and aromatic plants; (2) "Central Seed Testing Laboratory" means the Central Seed Testing Laboratory established or declared as such under sub-section (1) of section 32; (3) "Certification Agency" means an agency established under section 26 or accredited under section 27 or recognised under section 30; (4) "Chairperson" means the Chairperson of the Committee; (5) "Committee" means the Central Seed Committee constituted under sub-section (1) of section 3; (6) "container" means a

box, bottle, casket, tin, barrel, case, receptacle, sack, bag, wrapper or other thing in which any article or thing is placed or packed; (7) "dealer" means a person who carries on the business of buying and selling, exporting, or importing seed, and includes an agent of a dealer; (8) "essentially derived variety", in relation to a variety or an initial variety means a variety of seeds essentially derived from such initial variety when it- (a) is predominantly derived from such initial variety, or from a variety that itself is predominantly. derived from such initial variety, while retaining the expression of the essential characteristics that result from the genotype or combination of genotypes of such initial variety; (b) is clearly distinguished from such initial variety; and (c) conforms (except for the differences which result from the act of derivation) to such initial variety in the expression of the essential characteristics that result from the genotype or combination of genotypes of such initial variety; (9) "export" means taking out of India by land, sea or air. (10) "extant variety" means a variety available in India which(a) had been notified under section 5 of the Seeds Act, 1966 and remains as such on the date of commencement of this Act; or (b) is a farmers' variety as defined in clause (1) of section 2 of the Protection of Plant Varieties and Farmers' Rights Act, 2001; or (c) is a variety about which there is common knowledge; or (d) a variety other than a variety referred to in sub-clauses (a) to (c) and is in the public domain; (11) "farmer" means any person who cultivates crops either by cultivating the land himself or through any other person but does not include any individual, company, trader or dealer who engages in the procurement and sale of seeds on a commercial basis; (12) "horticulture nursery" means any place where horticulture plants are, in the regular course of business, produced or propagated and sold for transplantation; (13) "import" means bringing into India by land, sea or air. (14) "kind" means one or more related species or sub-species of crop plants each individually or collectively known by one common name such as cabbage, maize, paddy and wheat; (15) "member" means a member of the Committee; 54 of 1966 53 of 2001 (16) "misbranded" - A seed shall be deemed to be misbranded if- (i) it is a substitute for, or resembles in a manner likely to deceive, another variety of seed under the name of which it is sold, and is not plainly and conspicuously labelled so as to indicate its true nature; (ii) it is falsely stated to be the product of any place or country; (iii) it is sold by a name which belongs to another kind or variety of seed; (iv) false claims are made for it upon the label or otherwise; (v) when sold in a package which has been sealed or prepared by, or at the instance, of the dealer and which bears his name and address, the contents of each package are not conspicuously and correctly stated on the outside thereof within the limits of variability prescribed under this Act; (vi) the package containing it, or the label on the package bears any statement, design or device regarding the quality or the kind or variety of seed contained therein, which is false or misleading in any material particular or if the package is otherwise deceptive with respect to its contents; (vii) it is not registered in the manner required by or under this Act; (viii) the label contains any reference to registration other than the registration number; (ix) its label does not contain a warning or caution which may be necessary, and sufficient, if complied with, to protect human, animal and plant life and health or to avoid serious prejudice to the environment; (x) the package containing it or the label on the package bears the name of a fictitious individual or company as the dealer of the kind or variety; or (xi) it is not labelled in accordance with the requirements of this Act or the rules made thereunder. (17) "notification" means a notification published in the Official Gazette; (18) "prescribed" means prescribed by rules made under this Act; (19) "producer" means a person, group of persons, firm or organisation who grows or organizes the production of seeds; (20) "registered kind or variety", in relation to any seed, means any kind, or variety thereof, registered under section 13; (21) "Registration Sub-Committee" means the Registration Sub- Committee constituted under sub-section (1) of section 7; (22) "regulation" means a regulation made by the Committee under this Act; (23) "seed" means any type of living embryo or propagule capable of regeneration and giving rise to a plant of agriculture which is true to such type; (24) "Seed Analyst" means a Seed Analyst appointed under subsection. (1) of section 33; (25) "Seed Inspector" means a Seed Inspector appointed under sub-section (1) of section 34; (26) "seed processing" means the process by which seeds and planting materials are dried, threshed, shelled, ginned or delinted (in cotton), cleaned, graded or treated; (27) "spurious seed" means any seed which is not genuine or true to type; (28) "State Government" ,in relation to a Union territory, means the administrator thereof; (29) "State Seed Testing Laboratory", in relation to any State, means the State Seed Laboratory established or declared as such under sub-section (2) of section 32 for that State; (30) "transgenic variety" means seed or planting material synthesized or developed by modifying or altering the genetic composition by means of genetic engineering; (31) "variety" means a plant grouping except micro-organism within a single botanical taxon of the lowest known rank, which can be- (a) defined by the expression of the characteristics resulting from a given genotype of that plant grouping; (b) distinguished from any other plant grouping by expression of at least one of the said characteristics; and

(c) considered as a unit with regard to its suitability for being propagated, which remains unchanged after such propagation, and includes propagating material of such variety, extant variety, transgenic variety, farmers' variety and essentially derived variety.

3. (1) The Central Government shall, by notification, constitute, for the purpose of this Act, a Committee to be called the Central Seed Committee. (2) The headquarters of the Central Seed Committee shall be at New Delhi. Composition of the Committee.

4. (1) The Committee shall consist of a Chairperson, members, exofficio and other members, to be nominated by the Central Government. (2) The Secretary to the Government of India in the Department of Agriculture and Co-operation, Ministry of Agriculture, shall be Chairperson, ex officio. (3) The Committee shall consist of the following ex officio members, namely:- (i) the Agriculture Commissioner, Department of Agriculture and Co-operation, Government of India; (ii) the Deputy Director General (Crop Sciences), Indian Council of Agricultural Research; (iii) the Deputy Director General (Horticulture), Indian Council of Agricultural Research; (iv) the Joint Secretary in charge of seeds in the Department of Agriculture and Co-operation, Government of India; (v) the Horticulture Commissioner, Department of Agriculture and Co-operation, Government of India; (vi) a representative of the Department of Bio-technology, Government of India, not below the rank of Joint Secretary to the Government of India. (vii) a representative of the Ministry of Environment and Forests, Government of India, not below the rank of Joint Secretary to the Government of India. (4) The Committee shall consist of the following other members to be nominated by the Central Government, namely:- (i) the Secretary (Agriculture) from five States, one each from three out of the five geographical zones of the country as mentioned in the Schedule on rotation basis; (ii) Director, State Seed Certification Agency from one State which is not represented under clause (i); (iii) Managing Director, State Seeds Corporation, from one State which is not represented under clause. (i) or clause (ii); (iv) two representatives of farmers; (v) two representatives of seed industry; (vi) two specialists or experts in the field of seed development. (5) The Committee may associate with it, in such manner, on such terms and for such purposes as it may deem fit, any person whose assistance or advice it may desire in complying with any of the provisions of this Act, and a person so associated shall have the right to take part in the discussion of the Committee relevant to the purposes for which he has been associated, but shall not have the right to vote and shall be entitled to receive such

allowances or fees as may be fixed by the Central Government. (6) A Member nominated under sub-section (5) shall, unless his seat becomes vacant earlier by resignation, death or otherwise, be entitled to hold office for two years from the date of his nomination but shall be eligible for re-nomination provided that the said member shall hold office only for so long as he holds the appointment by virtue of which his nomination was made. (7) Save as otherwise provided, the terms and conditions of appointment of the members shall be such as may be prescribed. (8) A member other than an ex officio member may resign his office by giving notice in writing to the Central Government and on such resignation being accepted, he shall be deemed to have vacated his office. (9) A person shall be disqualified for being nominated or appointed as a member if he- (i) has been convicted and sentenced to imprisonment for an offence which, in the opinion of the Central Government, involves moral turpitude; or (ii) is an un-discharged insolvent; or (iii) is of unsound mind and stands so declared by a competent court. (10) No act or proceeding of the Committee shall become invalid merely by reason of – (i) any vacancy therein, or any defect in the constitution thereof; or (ii) any defect in the appointment of a person acting as the Chairperson or a member of the Committee; or (iii) any irregularity in the procedure of the Committee not affecting the merits of the case. (11) The Central Government may, at any time, remove from office any member other than member, ex-officio after giving him a reasonable opportunity of showing cause against the proposed removal. Powers and functions of the Committee.

5. The Committee shall be responsible for and shall have all the powers for the effective implementation of this Act and shall advise the Central Government and the State Governments on matters relating to- (a) seed programming and planning; (b) seed development and production; (c) export and import of seeds; (d) standards for registration, certification and seed testing; (e) seed registration and its enforcement; (f) such other matters as may be prescribed. Powers of Committee to specify minimum limits of germination, purity, seed health, etc.

6. The Committee may, by notification, specify– (a) the minimum limits of germination, genetic and physical purity, and seed health, with respect to any seed of any kind of variety; (b) the mark or label on the packet or container to indicate that such seed conforms to the minimum limits of germination, genetic and physical purity, and seed health specified under clause (a), and other particulars, such as expected performance of the seed in accordance with the information provided

by the producer under section 14 which such mark or label may contain. Registration and other Sub- Committees of the Committee and their functions.

7. (1) The Committee shall constitute a Sub-Committee to be called the Registration Sub-Committee consisting of a Chairman and such number of other members, to assist him in the discharge of the functions of the Committee, as may be prescribed. (2) It shall be the duty of the Registration Sub-Committee- (a) to register seeds of varieties after scrutinizing their claims as made in the application in such manner as may be prescribed; (b) to perform such other functions as are assigned to it by the Committee. (3) The Committee may appoint as many other Sub-Committees including a Sub-Committee on Seed Certification as it deems fit consisting wholly of the members of the Committee or wholly of other persons or partly of members of the Committee and partly of other persons as it thinks fit to exercise such powers and perform such duties as may be delegated to them. Procedure of the Committee and its Sub- Committees.

8. The Committee may, subject to the previous approval of the Central Government, make regulations for the purpose of regulating its own procedure and the procedure of any Sub-Committee thereof. Secretary and other officers of the Committee.

9. The Central Government shall – (a) appoint a person to be the Secretary of the Committee; and (b) provide the Committee with such technical and other officers and employees as may be necessary for the efficient performance of the functions of the Committee under this Act. Meetings of the Committee.

10. (1) The Committee shall meet as and when necessary at such time and place and shall observe such procedure in regard to transaction of business at its meetings (including the quorum at meetings) as may be provided by regulations. (2) The Chairperson or, in his absence, the Agricultural Commissioner or, in the absence of both the Chairperson and the Agricultural Commissioner, any member chosen by the members present from amongst themselves, shall preside at a meeting of the Committee. (3) All questions at a meeting of the Committee shall be decided by a majority of votes of the members present and voting and in the case of an equality of votes, the Chairperson or, in his absence, the Agricultural Commissioner or, in the absence of both the Chairperson and the Agricultural Commissioner the person presiding shall have and exercise a second or casting vote. State Seed Committee.

11. Every State Government shall establish a State Seed Committee to – (a) advise the Committee on registration of regional or local seeds of any kind or variety; (b) advise the State Government on registration of seed producing units, seed processing units, seed dealers and horticulture nurseries; (c) maintain, in each district, a list of seed dealers, seed producers, seed processing units and horticulture nurseries; (d) seek information from persons engaged in the production, supply, distribution, trade or commerce in seeds of any kind or variety regarding stocks, prices, sales and other information in the manner as may be prescribed; (e) advise the State Government and the Committee on all matters arising out of the administration and implementation of this Act; and (f) carry out other functions assigned to, by, or under this Act.

Chapter III Registration of Kinds and Varieties of Seeds, etc:
Maintenance of National Register of seeds of kinds and varieties:

12. (1) For the purposes of this Act, a register of all kinds and varieties of seed to be called the National Register of Seeds shall be kept by the Registration Sub-Committee wherein all specifications, as may be prescribed, shall be maintained. (2) Subject to the directions of the Committee, the Register shall be kept under the control and management of the Registration Sub- Committee. (3) The Registration Sub-Committee shall, within such intervals and in such manner as it thinks appropriate, publish the list of kinds and varieties of seed which have been registered during that interval. Registration of seeds of any kind or variety.

13. (1) No seed of any kind or variety shall, for the purpose of sowing or planting by any person, be sold unless such seed is registered under sub-section (2) by the Registration Sub-Committee in such manner as may be prescribed. (2) Subject to the provisions of sections 14 and 15, the Registration Sub-Committee may register or refuse to register any kind or variety of seed on the basis of information furnished by the producer who develops the variety on the results of multi-locational trials for such period as may be prescribed to establish the performance of that seed: Provided that no application for registration shall be refused under this sub-section unless the applicant has been given an opportunity to represent his case: Provided further that seeds of any kind or variety in respect of which a valid registration exist on the date of commencement of this Act are not required to be registered again under this section on basis of the information on the results of multi-location trials. (3) The Registration Sub-Committee may grant provisional registration to the varieties of seeds, which are available in the market on the date of commencement of this Act. (4) Registration made under this Act shall be valid for a period of fifteen years in the case of annual

and biennial crops, and eighteen years for long duration perennials. (5) At the expiry of the period granted under sub-section (4), the kind or variety of seed may be re-registered for a like period by the Registration Sub-Committee on the basis of information furnished by the producer on the results of such trials as may be prescribed under sub-section (2) to re-establish the performance of the kind or variety of seed. (6) The Registration Sub-Committee shall have the power to issue such directions to protect the interests of a producer against any abusive act committed by any third party during the period between the date of filing of application for registration and the date of decision by the Committee on such application. Procedure for registration.

14. (1) Every application for registration under sub-section (2) section 13 shall be made in such form and contain such particulars and be accompanied by such fees as may be prescribed. (2) On receipt of any such application for the registration of a kind or variety of seed, the Registration Sub-Committee may, after such enquiry as it deems fit and after satisfying itself that the kind or variety of seed to which the application relates conforms to the claims made by the importer or by the seller, as the case may be, as regards the efficacy of the kind or variety of seed and its safety to human beings and animals, register the kind or variety, as the case may be, of the seed on such conditions as may be specified by it and allot a registration number thereto and issue a certificate of registration. (3) The Registration Sub-Committee may, having regard to the efficacy of the seeds and its safety to human beings and animals, vary the conditions subject to which a certificate of registration has been granted and may, for that purpose, require the certificate holder by notice in writing to deliver the certificate to it within such time as may be specified in the notice. Special provision for registration of transgenic varieties.

15. (1) Notwithstanding anything contained in section 14, no seed of any transgenic variety shall be registered unless the applicant has obtained clearance in respect of the same as required by or under the provisions of the Environment (Protection) Act, 1986: Provided that the Registration Sub-Committee may, subject to clearance under the said Act, grant provisional registration, for a period not exceeding two years on the basis of information furnished by the producer on the results of multi-locational trials conducted in the prescribed manner. (2) Save as otherwise provided in sub-section (1), the form and manner in which and procedure for registration of transgenic variety of seed and the fee payable thereto shall be the same as applicable in case of registration under section 14. 29 of 1986 Cancellation of registration of seeds of kinds and varieties.

16. (1) The Registration Sub-Committee may cancel any registration granted under section 13 or section 15 or any one or more of the following grounds, namely:- (a) that the holder of the certificate has violated any of the terms and conditions of the registration; or (b) that the registration has been obtained by misrepresentation or concealment of essential data; or (c) that the variety is not performing in accordance with the information furnished by the producer under sub-section (2) of section 13 or has become obsolete or has outlived its utility and be made available for sale in the market in the public interest; or (d) that prevention of commercial exploitation of such variety of seeds is necessary. (i) in the public interest; (ii) to protect public order or public morality; or (iii) to protect human beings, animal and plant life and health to avoid serious prejudice to the environment. (2) No order of cancellation of registration under this section shall be made unless the holder thereof or the affected person concerned has been given a reasonable opportunity of showing cause in respect of the grounds for such cancellation. Notification of cancellation of registration of seeds of kinds and varieties.
17. The Registration Sub-Committee shall notify the cancellation of registration of any kind or variety of seed made under section 13 or any registration made under section 15 in the Official Gazette Exclusion of certain kinds or varieties of seed from registration. Evaluation of performance.
18. (1) Notwithstanding anything contained in this Act, no registration of any kind or variety of seed shall be made under this Act, if prevention of commercial exploitation of such kind or variety is necessary to protect public order or public morality or human, animal or plant life and health, or to avoid serious prejudice to the environment. (2) A kind or variety of seed containing any technology, which is harmful, or potentially harmful, shall not be registered. Explanation.- For the purposes of this sub-section, the expression "technology" includes genetic use restriction technology and terminator technology.
19. The Committee may, for conducting trials to assess the performance, accredit centres of the Indian Council of Agricultural Research, State Agricultural Universities and such other organizations fulfilling the eligibility requirements as may be prescribed, to conduct trials to evaluate the performance of any kind or variety of seed. Compensation to farmer.
20. Where the seed of any registered kind or variety is sold to a farmer, the producer, distributor or vendor, as the case may be, shall disclose the expected performance of such kind or variety to the farmer under given conditions, and if, such registered seed fails to provide the expected

performance under such given conditions, the farmer may claim compensation from the producer, dealer, distributor or vendor under the Consumer Protection Act, 1986. Seed producers and seed processing units to be registered.

21. (1) No producer shall grow or organize the production of seed unless he is registered as such by the State Government under this Act. (2) No person shall maintain a seed processing unit unless such unit is registered by the State Government under this Act. (3) The State Government shall register a producer or seed processing unit if he or it meets the specifications prescribed by the Central Government in terms of infrastructure, equipment and qualified manpower. (4) Every application for registration under sub-section (3) shall be made in such form and manner and shall be accompanied by such fee as may be prescribed. (5) The State Government may, after making such enquiry and subject to such conditions as it thinks fit, grant a certificate for maintaining a seed producing or a seed processing unit in such form as may be prescribed. (6) Every seed producing unit and every seed processing unit shall furnish periodic returns on the quantity of seeds of different kinds or varieties produced or processed by it to the Seed Certification Agency in such form and at such time as may be prescribed. (7) The State Government may, after giving the holder of certificate of registration under sub-section (1), or sub-section (2), as the case may be, suspend or cancel the registration if ¾ (a) such registration has been obtained by misrepresentation as to a material particular relating to the specification in terms of infrastructure, equipment or availability of qualified manpower; or (b) any of the provisions of this Act or the rules made thereunder has been contravened. Seed dealers to be registered.

22. (1) Every person who desires to carry on the business of selling, keeping for sale, offering to sell, bartering, import or export or otherwise supply any seed by himself, or by any other person on his behalf shall obtain a registration certificate as a dealer in seeds from the State Government. (2) Every applicant for dealership under sub-section (1) shall be required to furnish information about seed stocks, sales and other related information as may be prescribed. (3) Even application for registration under sub-section(1) shall be made in such form and manner and shall be accompanied by such fee as may be prescribed. (4) The State Government may, after making such enquiry and subject to such conditions as it thinks fit, grant a certificate of registration as a dealer in seeds in such form as may be prescribed. (5) Every dealer registered under this section shall furnish to the State Government such information

and returns regarding seed stocks, seed lots, expiry date of seed lots and other related information as may be prescribed. (6) The State Government may, after giving the dealer an opportunity of being heard, suspend or cancel a certificate granted under this Act if- (a) such registration had been obtained by misrepresentation of any material fact; (b) contravenes any of the provisions of this Act or the rules made thereunder. Horticulture nursery to be registered.

23. (1) No person shall conduct or carry on the business of horticulture nursery for any of the purposes of this Act unless such nursery is registered with the State Government. 24. Every application for registration under sub-section (1) shall be made in such form and contain such particulars and shall be accompanied by such fee as may be prescribed. Duties of registration holders of horticulture nursery.

24. Every person who is a holder of a registration of a horticulture nursery under section 23 shall- (a) keep a complete record of the origin or source of every planting material and performance record of mother trees in the nursery; (b) keep a layout plan showing the position of the root-stocks and scions used in raising the horticulture plants; (c) keep a performance record of the mother trees in the nursery; (d) keep the nursery plants as well as the parent trees used for the production or propagation of horticulture plants free from infectious or contagious insects, pests or diseases affecting plants; and (e) furnish such information to the State Government on the production, stocks, sales and prices of planting material in the nursery as may be prescribed.

Chapter IV Regulation of Sale of Seed and Seed Certification Agencies: Regulation of sale of seeds of registered kinds and varieties.

25. No person shall himself, or by any other person on his behalf, carry on the business of selling, keeping for sale, offering to sell, bartering, import or export or otherwise supply any kind of seed of any registered kind or variety unless- (a) such seed is identifiable as to its kind or variety; (b) such seed conforms to the minimum limit of germination and genetic, physical purity, maximum seed health specified under clause (a) of section 6; © the container of such seed bears in the prescribed manner, the mark or label bearing the correct particulars thereof, specified under clause (b) of section 6; (d) the container of such seed, in the case of transgenic varieties, bears a declaration to this effect; and 25. He complies with such other requirements as may be prescribed. State Seed Certification Agency.

26. The Committee may, in consultation with the State Government, by notification, establish a State Seed Certification Agency for the State to carry out the functions entrusted to the State Seed Certification

Agency by or under this Act: Accreditation of Seed Certification Agencies.

27. (1) The Committee may in consultation with the State Government and the State Seed Committee, accredit – (a) organizations to carry out certification, on the fulfillment of such criteria, as may be prescribed, or (b) individuals or seed producing organisations to carry out selfcertification, in such manner as may be prescribed. (2) The accredited individuals and seed producing organisations shall be subject to such inspection and control of the Committee, the concerned State Government and State Seed Certification Agency, as may be prescribed. (3) The accreditation may be withdrawn by the Committee, for reasons to be recorded in writing and after giving to the concerned organization or individual, as the case may be, a reasonable opportunity of being heard. Grant of certificate by the State Seed Certification Agency.

28. (1) Any person selling, keeping for sale, offering to sell, bartering or otherwise supplying any seed of any registered kind or variety may, if he desires to have such seed certified by the State Seed Certification Agency, apply to that Agency for the grant of a certificate for the purpose. (2) Every application under sub-section (1) shall be made in such form, shall contain such particulars and shall be accompanied by such fee as may be prescribed. (3) On receipt of an application under sub-section (1), the State Seed Certification Agency may, after such enquiry as it thinks fit and after satisfying itself that the seed to which the application relates conforms to the prescribed standards, grant a certificate in such form and on such conditions as may be prescribed: Provided that such standards shall not be lower than the minimum limit of germination, genetic and physical purity specified for that seed under clause (a) of section 6. Revocation of certificate.

29. If the State Seed Certification Agency is satisfied, either on a reference made to it in this behalf or otherwise, that- (a) the certificate granted by it under section 28 has been obtained by misrepresentation as to an essential fact; or (b) the holder of the certificate has, without reasonable cause, failed to comply with the conditions subject to which the certificate has been granted or has contravened any of the provisions of this Act or the rules made thereunder, then, without prejudice to any other penalty to which the holder of the certificate may be liable under this Act, the State Seed Certification Agency may, after giving the holder of the certificate an opportunity of showing cause, revoke the certificate. Recognition of seed certification agencies in foreign countries.

30. The Central Government may, on the recommendation of the Committee and by notification, recognise any seed certification agency established in any foreign country, for the purposes of this Act.

Chapter V Appeals Appeals:

31. (1) Any person aggrieved by a decision of the Registration Sub- Committee under section 14, section 16 or section 27 or of the State Seed Certification Agency under section 28 or section 29 may, within thirty days from the date on which the decision is communicated to him prefer an appeal to such authority (hereinafter referred to as the appellate authority) as the Central Government may think fit to constitute: Provided that the appellate authority may entertain an appeal after the expiry of the said period of thirty days if it is satisfied that the appellant was prevented by sufficient cause from filing the appeal in time. (2) An appellate authority shall consist of a single person or three persons as the Central Government may think fit, to be appointed by that Government. (3) The form and manner in which an appeal may be preferred under sub-section (1), the fee payable for such appeal and the procedure to be followed by the appellate authority shall be such as may be prescribed. (4) On receipt of an appeal preferred under sub-section (1), the appellate authority shall, after giving the appellant and the other party an opportunity of being heard, dispose of the appeal as expeditiously as possible.

Chapter VI Seed Analysis and Seed Testing: Central and State Seed Testing Laboratories.

32. (1) The Central Government may, by notification, establish a Central Seed Testing Laboratory or declare any seed-testing laboratory as the Central Seed Testing Laboratory to carry out the functions entrusted to the Central Seed Testing Laboratory by or under this Act in the prescribed manner. (2) The State Government may, in consultation with the Committee, and by notification, establish one or more State Seed Testing Laboratories or declare any seed testing laboratory in the Government or non-Government sector as a State Seed Testing Laboratory where analysis of seed of any kind or variety shall be carried out under this Act in the prescribed manner. (3) Every Seed Testing Laboratory referred to in sub-section (1) shall have as many Seed Analysts as the Central Government may consider necessary. (4) Every Seed Testing Laboratory referred to in sub-section (2) shall have as many Seed Analysts as the State Government may consider necessary. Seed Analysts.
33. (1) In case of the Central Seed Laboratory, the Central Government and in other cases the State Government may, by notification, appoint such persons as the concerned Government thinks fit and having the prescribed qualifications to be Seed Analysts and define the local limits of their jurisdiction. (2) Every Central Seed Testing Laboratory established or declared under sub-section (1) of section 32 and every State Seed

Testing Laboratory established or declared under sub-section (2) of that section shall have as many Seed Analysts as the Central Government or the State Government, as the case may be, specify. Seed Inspectors.

34. (1) The State Government may, by notification, appoint such persons as it thinks fit, having the prescribed qualifications, to be Seed Inspectors and define the areas within which they shall exercise jurisdiction. (2) Every Seed Inspector shall be subordinate to such authority as the State Government may specify in this behalf. Powers of Seed Inspectors.

35. (1) The Seed Inspector may- (a) take samples of any seed of any kind or variety from- (i) any person selling such seed; or (ii) any person who is in the course of conveying, delivering or preparing to deliver such seed to a purchaser or a consignee; or (iii) a purchaser or a consignee after delivery of such seed to him; (b) send such sample for analysis to the Seed Analyst of the area within which such sample has been taken; (c) enter and search, at all reasonable times, with such assistance, if any, as he considers necessary, any place in which he has reason to believe that an offence under this Act has been or is being committed and order in writing the person in possession of any seed in respect of which the offence has been or is being committed, not to dispose of any stock of such seed for a specific period not exceeding thirty days or, unless the alleged offence is such that the defect may be removed by the possessor of the seed, seize the stock of such seed; (d) examine any record, register, document or any other material object found in any place mentioned in clause (c) and seize the same if he has reason to believe that it may furnish evidence of the commission of an offence punishable under this Act; and (e) exercise such other powers as may be necessary for carrying out the purposes of this Act or any rule or regulation made thereunder. (2) The power conferred by this section includes the power to break-open any container in which any seed of any kind or variety may be contained or to break-open the door of any premises where any such seed may be kept for sale: Provided that the power to break-open the door shall be exercised only after the owner or any other person in occupation of the premises, if he is present therein, refuses to open the door on being called upon to do so. (3) Where the Seed Inspector takes any action under clause (a) of sub-section (1), he shall, as far as possible, call not less than two independent and respectable persons to be present at the time when such action is taken and take their signatures on a memorandum to be prepared in such form and manner as may be prescribed. (4) The provisions of the Code of Criminal Procedure, 1973, or in relation to the State of Jammu and Kashmir, the provisions of any corresponding law in force in that State,

shall, so far as may be, apply to any search or seizure under this section as they apply to any search or seizure made under the authority of a warrant issued under section 94 of the said Code, or, as the case may be, under the corresponding provisions of the said law.

Chapter VII Export and Import of Seeds Import of seeds:

36. (1) All import of seeds – (a) shall be subject to the provisions of the Plant Quarantine (Regulation of Import into India) Order, 2003, or any corresponding order made under section 3 of the Destructive Insects and Pests Act, 1914; (b) shall conform to minimum limits of germination, genetic and physical purity, and seed health as prescribed under section 6; and (c) shall be subject to registration as may be granted on the basis of information furnished by the importer on the results of multilocational trials for such period as may be prescribed to establish the performance. (2) The Central Government may, by notification, permit to import an unregistered variety in such quantity and subject to fulfilling such conditions as may be specified in that notification for research purposes. 2 of 1914. Export of seeds.
37. The Central Government may, on the advice of the Committee, restrict, by notification, the export of seeds of any kind or variety if it is deemed that such export may adversely affect the food security of the country, or if it is felt that the reasonable requirements of the public will not be met, or on such other grounds as may be prescribed.

Chapter VIII Offences and Punishment Offences and punishment:

38. (1) If any person – (a) contravenes any provision of this Act or any rule made thereunder; or (b) imports, sells, stocks or exhibits for sale or barter or otherwise supplies any seed of any kind or variety deemed to be misbranded ; or (c) imports, sells, stocks or exhibits for sale or barter, or otherwise supplies any seed of any kind or variety without a certificate of registration; or (d) obstructs the Committee, Registration Sub-Committee or Seed Certification Agency or Seed Inspector or Seed Analyst or any other authority appointed or duly empowered under this Act in the exercise of its powers or discharge of their duties under this Act or the rules made thereunder, he shall, on conviction, be punishable – with fine which shall not be less than five thousand rupees but which may extend to twenty five thousand rupees. (2) If any person sells any seed which does not conform to the standards of physical purity, germination or health or does not maintain any records required to be maintained under this Act or the rules made thereunder he shall, on conviction, be punishable with fine which shall not be less than five thousand rupees but which may extend to twenty- five thousand rupees. (3) If any person furnishes any false information relating to the standards of genetic purity, misbrands any seed or supplies any spurious seed or

spurious transgenic variety, sells any non-registered seeds he shall, on conviction be punishable with imprisonment for a term which may extend to six months or with fine which may extend to fifty thousand rupees or with both. Forfeiture of property.

39. When any person has been convicted under this Act for the contravention of any of the provisions of this Act or the rules made thereunder, the seed in respect of which the contravention has been committed shall be forfeited to the Central Government. Offences by companies.

40. (1) Where an offence under this Act has been committed by a company, every person who at the time the offence was committed was in charge of, and was responsible to the company for the conduct of the business of the company, as well as the company, shall be deemed to be guilty of the offence and shall be liable to be proceeded against and punished accordingly: Provided that nothing contained in this sub-section shall render any such person liable to any punishment under this Act if he proves that the offence was committed without his knowledge and that he exercised all due diligence to prevent the commission of such offence. (2) Notwithstanding anything contained in sub-section (1), where an offence under this Act has been committed by a company and it is proved that the offence has been committed with the consent or connivance of, or is attributable to any neglect on the part of, any director, manager, secretary or other officer of the company, such director, manager, secretary or other officer shall also be deemed to be guilty of that offence and shall be liable to be proceeded against and punished accordingly. Explanation. – For the purpose of this section,- (a) “company” means any body corporate and includes a firm or other association of individuals; and (b) “director”, in relation to a firm, means a partner in the firm.

Chapter IX Power of Central Government: Power of Central Government to give directions to the State Governments.

41. The Central Government may give such directions to any State Governments as may appear to the Central Government to be necessary for carrying into execution in the State any of the provisions of this Act or of any rule made there under. Power of Central Government to issue directions to the Committee.

42. (1) Without prejudice to the foregoing provisions of this Act, the Committee shall, in the discharge of its functions and the performance duties under this Act, be bound by such directions on questions of policy as the Central Government may give in writing to it from time to time. (2) The decision of the Central Government whether a question is one of policy or not shall be final. Exemption from registration.

43. (1) Nothing in this Act shall restrict the right of the farmer to save, use, exchange, share or sell his farm seeds and planting material, except that he shall not sell such seed or planting material under a brand name or which does not conform to the minimum limit of germination, physical purity, genetic purity prescribed under clause (a) or clause (b) of section 6. (2) The Central Government may, by notification, and subject to conditions, if any, as it may specify therein, exempt from all or any of the provisions of this Act or the rules made thereunder, any educational, scientific or research or extension organization.

Chapter X Miscellaneous Protection of action taken in good faith:

44. No suit, prosecution or other legal proceeding shall lie against the Government or any person for anything which is in good faith done or intended to be done under this Act. Power to remove difficulties.

45. (1) If any difficulty arises in giving effect to the provisions of this Act, the Central Government may, by order published in the Official Gazette, make such provisions not inconsistent with the provisions of this Act as may appear to be necessary for removing the difficulty: Provided that no order shall be made under this section after the expiry of two years from the date of commencement of this Act. (2) Every order made under sub-section (1) shall be laid before each House of Parliament. Power of Central Government to make rules.

46. (1) The Central Government may by notification, make rules to carry out the provisions of this Act. (2) In particular and without prejudice to the generality of the foregoing power, such rules may provide for all or any of the following matters, namely:- (a) the terms and conditions of appointment of members of the Committee under sub-section (7) of section 4; (b) the matters to be prescribed under clause (f) of section 5; (c) the number of other members who shall assist the Chairperson of the Registration Sub-Committee to discharge its functions under sub-section (1) of section 7; (d) the manner of scrutinizing the claims as made in the applications under clause (a) of sub-section (2) of section 7; (e) the manner of seeking information by a State Seed Committee under Clause (d) of Section 11; (f) the specifications which shall be maintained in the National Register of Seeds of kinds or varieties under sub-section (1) of section 12; (g) the manner of registration of seed of any kind or variety under sub-section (1) of section 13; (h) the period which required to establish performance of seeds on the results of multi-locational trials conducted under subsection (2) of section 13; (i) the form of application and the particulars which should be furnished, and the fee which should accompany, such application under sub-section (1) of section 14; (j) the eligibility requirement which an

organization shall fulfil for accreditation under section 19; (k) the specification required to be fulfilled for registration as a producer or seed producing unit under sub-section (3) of section 21; (l) the form and manner in which an application for registration under sub-section (3) of section 21 shall be made and the fee with which such application shall be accompanied under subsection (4) of said section ; (m) the form in which a certificate for maintaining a seed producing or seed processing unit may be granted under sub-section (5) of section 21; (n) the form in which and time within which periodic returns shall be filled under sub-section (6) of section 21; (o) the information which an application for dealership in seeds shall be furnished under sub-section (2) of section 22; (p) the form and manner in which an application for registration as seed dealer under sub-section (1) of section 22 shall be made and the fee which shall accompany such application under subsection (3) of that section; (q) the form in which a certificate of registration as a dealer in seeds shall be granted under sub-section (4) of section 22; (r) the information and return which a registered dealer shall furnish to the State Government under sub-section (5) of section 22;; (s) the form in which an application for registration of a horticulture nursery shall be made, the particulars which such application shall contain and fee which shall accompany such application under sub-section (2) of section 23; (t) the information on production, stocks, sales and prices of planting material in a nursery shall be furnished to the State Government under section 24; (u) the manner in which the container of seeds shall bear the mark or label under clause (c) of section 25. (v) the requirement which a person carrying on business of selling, etc. of any registered kind or variety of seeds shall comply with under clause (e) of section 25; (w) the criteria to be fulfilled under clause (a) and the manner of carrying out self-certification under clause (b) of sub-section(1) of section 27; (x) the inspection and control which Committee, the concerned State Government and the State Seeds Certification Agency shall carryout or have on an accredited individuals and seed producing organizations under sub-section (2) of section 27; (y) the form of application and the particulars to be furnished in such application and the fee which shall accompany such application under sub-section (2) of section 28; (z) the form in which and the conditions subject to which a certificate shall be granted under sub-section (3) of section 28; (za) the form and manner in which an appeal shall be preferred and the fee which such appeal shall accompany and the procedure which the Appellate Authority shall follow under sub-section (3) of section 31; (zb) the manner in which a Central Seed Testing Laboratory established or declared under sub-section (1) of section 32 shall carry out its functions; (zc) the manner of carrying out analysis of seeds shall be made under sub-section (2) of section 32; (zd) the qualifications which a person to be appointed as Seed Analysts shall possess under sub-section (1) of section 33; (ze) the qualifications which a person to be appointed as Seed Inspector shall possess under sub-section (1) of section 34; (zf) the form and manner in which the memorandum shall be prepared under sub-section (3) of section 35; (zg) the grounds on which the Central Government may restrict export of seeds under section 37; (zh) any other matter which is to be or may be prescribed. Power of Committee to make regulations.

47. (1) The Committee may, with the previous approval of the Central Government, by notification, make regulations not inconsistent with the provisions of this Act and the rules made thereunder, to provide for all matters for which provision is necessary or expedient for the purpose of giving effect to the provisions of this Act. (2) In particular and without prejudice to the generality of the foregoing power, such regulations may provide for all or any of the following matters, namely:- (a) the procedure for conduct of business to be transacted by the Committee or any Sub-Committee thereof under section 8; (b) the procedure in regard to transaction of business at meetings of the Committee (including the quorum at meetings) under sub-section (1) of section 10. (c) any other matter in respect of which regulations are to be or may be made Rules and regulations to be laid before Parliament.

48. Every rule and every regulation made under this Act shall be laid as soon as may be after it is made, before each House of Parliament, while it is in session, for a total period of thirty days which may be comprised in one session or in two or more successive sessions, and if, before the expiry of the session immediately following the session or the successive sessions aforesaid, both Houses agree in making any modification in the rule or regulation or both Houses agree that the rule or regulation should not be made, the rule or regulation shall, thereafter, have effect only in such modified form or be of no effect, as the case may be; so, however, that any such modification or annulment shall be without prejudice to the validity of anything previously done under that rule or regulation. Repeal and savings.

49. (1) The Seeds Act, 1966 is hereby repealed. (2) Without prejudice to the provisions contained in the General Clauses, 1897, with respect to repeals, no such repeal shall affect- (a) the previous operation of the law so repealed or anything duly done or suffered thereunder; or (b) any right, privilege, obligation or liability acquired, accrued or incurred under the law so repealed; or (c) any penalty, forfeiture or punishment

incurred in respect of any offence committed against the Act so repealed; or (d) any investigation, proceeding, legal proceeding or remedy in respect of any such right, privilege, obligation, liability, penalty, forfeiture or punishment as aforesaid; and any such investigation, proceedings, legal proceeding or remedy may be instituted, continued or enforced; any such penalty forfeiture or punishment may be imposed as if this Act had not been passed: Provided further that, subject to the first proviso and any saving provisions made elsewhere in this Act anything done, any action taken, any rule made, any notifications or orders issued under 54 of 1966 10 of 1897 the provisions of the Act so repealed shall, in so far as they are not inconsistent with the provisions of this Act, be deemed to have been done, taken, made or issued under the corresponding provisions of this Act, and shall continue to be in force accordingly, unless and until expressly or implied repealed by any thing done, action taken, rules made or, notification or orders issued under this Act. (2) Notwithstanding such repeals any kind or variety of seeds that has been notified under the law as so repealed shall be deemed to have been registered under this Act, and any seed certification agency established under section 18 of the Seeds Act, 1966 shall be deemed to have been established or recognised, as the case may be, under this Act.

3

Seed Evolution and Seed Development

Evolution of Seed

The evolution of plants has resulted in increasing levels of complexity, from the earliest algal mats, through bryophytes, lycopods, ferns to the complex gymnosperms and angiosperms of today. While the groups which appeared earlier continue to thrive, especially in the environments in which they evolved, each new grade of organisation has eventually become more "successful" than its predecessors by most measures.

1. Evidence suggests that an algal scum formed on the land 1,200 million years ago
2. To thrive and to avoid extinction, plant are made mechanisms and evolved seed plant during 200 million years ago
3. The latest major group of plants to evolve were the grasses, 40 million years ago
4. The grasses, as well as many other groups, evolved new mechanisms of metabolism to survive the low CO_2 and warm, dry conditions of the tropics over the last 10 million years.

Seed Development, Maturation and Seed Structure

A true seed is defined as a fertilized mature ovule consisting of embryo, stored food material and protective coats. The important events involved in seed development and maturation include

1. Pollination
2. Fertilization

3. Development of the fertilized ovule by cell division
4. Accumulation of reserve food material
5. Loss of moisture content.

Pollination

The mature anthers dehisce and release pollen -grains (haploid microspores). When pollen grains are transferred from an anther to the stigma of the same flower the process is called self-pollination or autogamy. If they are transferred to the stigma of another flower, cross-pollination or allogamy is said to have occurred. Self-pollination occurs in those plants where bisexual flowers achieve anther dehiscence and stigma receptivity simultaneously called as chasmogamy. The majority of angiosperms bear chasmogamous flowers.

In some plants, flowers do not open before pollination such flowers are called cleistogamous, and this is the most efficient floral adaptation for promoting self-pollination. Cross-pollination is ensured in plants which bear unisexual flowers. In bisexual flowers also self-pollination may be prevented by

a. *Self-sterility:* inability to produce viable pollen Eg:Sunflower
b. *Dichogamy:* maturation of male and female organs at different times Eg:Bajra
c. *Herkogamy:* where the structure of male and female sex organs proves a barrier to self pollination Eg: Lucernae
d. *Heterostyly:* where flowers are of different types depending on the length of the style and stigma and pollination occurs only between 2 dissimilar types Eg: Brassica
e. *Self incompatibility:* Inability to viable pollen to fertilize ovule of same flower Eg: Cole crops

Self-pollinated crops: wheat, rice, barely, mungbean and cowpea. Cross pollinated Crops: Maize, rye, carrot, cauliflower and onion. Often cross pollinated crops: cotton and pigeon pea where there may be 10-40 % cross pollination.

Agents bring about the dissemination of pollen grains

1. *Abiotic:* wind (anemophily) and water (hydrophily)
2. Biotic including insects (entomophily) and bats (cheiropterophily).

Fertilization

After landing on the stigma, the pollen grain germinates and pollen tube grows through the style. The surface of the stigma secretes substances, which may provide optimum conditions for pollen germination. The pollen tubes traversing the style pectinase which dissolves intercellular substances of the style tissue. After traversing the style, the pollen tube enters embryosac of the ovule. The embryosac consists of 8 cells. The end near the micropyle has the egg apparatus, which consists of egg cell and 2 synergids. There are 2 polar nuclei in the centre and the chalazal end has 3 antipodal cells.

In angiosperms, fertilization involves the participation of 2 male nuclei (double fertilization). One fuses with the egg nucleus to form the diploid zygote and the other with 2 polar nuclei to produce a triploid nucleus, which is the primary endosperm nucleus.

Seed Development

Embryo Development

The first division of the zygote is transverse in dicots and it results in a small apical cell and a large basal cell. Cell ca divides vertically forming 2 juxtaposed cells and cb undergoes a transverse division forming 2 superimposed cells. These results in a T-shaped, 4 celled proembryo. Cell ci divides transversely giving rise to n and n'. These 2 cells divide further resulting in a row of 3 or 4 cells, forming suspensor.

Cell m and its derivatives undergo vertical divisions forming a group of 4 to 6 cells. This group divides by oblique-perclinal wall forming a set of inner cells and a row of outer cells. The inner cells form the initials of the root apex and the outer cells form the root cap. The 2 cells formed as a result of the division of ca again divide vertically forming quadrant. Each cell of the quadrant divides transversely and thus an octant containing 2 tiers of cell l and p is formed.

The cells of the octant undergo vertical division resulting in a globular proembryo. Periclinal divisions occur in the peripheral cells of the globular proembryo that delimit an outer layer, the dermatogen. The tier l gives rise to cotyledons and shoot apex and l forms hypocotl-radicle axis. Certain deviations from the above pattern of embryo development are found in different plants. Different types of embryogeny are distinguished depending on the plane of division of the apical and the extent of contribution of the basal cell towards embryo development (in some plants cb remains undivided and does not take part in embryo development at all).

In monocotyledons, the cell cb remains undivided and develops into a haustorial of the suspension. Cell ca divides into 2 by a transverse division. The terminal cell of these 2 by repeated divisions in different planes gives rise to a single cotyledon. The embryo development in grasses is different from that of other monocotyledons. A dorsiventral symmetry is established as a result of the peculiar oblique position of cell walls early in the embryogeny. The single cotyledon is reduced to absorptive scutellum and additional structures like coleptile and coleorrhiza are formed.

Endosperm Development

There are 3 types of endosperm development (a) nuclear - where the endosperm nucleus undergoes several divisions prior to cell wall formation, e.g., wheat apple, squash, (b) cellular -in which there is no free nuclear phase, and (c) helobial where the free nuclear division is preceded, and is followed by cellularization as in some monocots. During the course of seed development, reserve food materials are accumulated in the endosperm from the adjacent tissues.

In endospermic dicot seeds, endosperms are retained as a permanent storage tissue. In non-endospermic dicot seeds, endosperm reserves are depleted and occluded by the developing embryo. The reserves are then reorganized in the cotyledons, which in turn act as the source of stored reserved food for embryo after germination. A part of the endosperm is depleted in cereals during embryo maturation and this lies as a layer between the starchy endosperm and scutellum.

Seed-Coat Development: Integument's of the ovule undergo marked reorganization and histological changes during maturation to form seed coats. In bitegmic ovules (which have 2 integument's), the seed coat may be derived from both the integument's or from the outer integument only; the inner integument may disintegrate.

Seed Structure and Functions

Seeds may be broadly classified as dicotyledons and monocotyledons, depending on the number of cotyledons. Dicotyledons seeds may be either non-endospermic (exalbuminous) e.g. chickpea, pea and bean or endospermic (albuminous) e.g., castorbean, fenugreek, etc., Monocotyledons seeds are mostly albuminous.

A typical non-endospermic dicot seed is made up of seed coat and embryo. The seed coat consists of 2 layers that may be united or free, the outer layer, which is hard and made of thick walled cells is called testa and the inner thin membranous layer is called tegument. The seed coat is of considerable importance because it is the only protective barrier for the embryo from the external environment.

1. The seed coat bears a scar called hilum, marking the point at which seed is attached to stalk.
2. The funicle or the stalk forms a ridge called raphe along the margin of the seed.
3. At one end of the hilum, there is a small hole called micropyle. There is an outgrowth below the hilum in leguminous seeds, which is called strophiole.

4. Certain other seeds (castorbean, nutmeg) have outgrowths called arials.
5. Arillar contents may important in attracting animals, which aid in seed dispersal.

The embryo consists of embyonic axis and 2 fleshy cotyledons. The axis includes embyonic root (radicle), hypocotyl to which 2 cotyledons are attached and plumule (shoot apex with first true leaves). The cotyledons of non-endospermic e.g., pea are bulky and account for over 90% of the mass of the seed.

Cotyledons of epigeal, non-endopermic species become leaf like and photosynthetic after germination. In endospermic dicot seeds, the endosperm is bulky and stores food reserves. In these cases, the cotyledons are small or haustorial in nature.

The nucleus of the ovary after fertilization becomes perisperm. The perisperm in the majority of seeds fails to pass beyond an incipient stage of development but in a few cases this tissue becomes the store for food reserves such as coffee.

Poaceae seeds are generally enclosed in one seeded fruit called caryopsis. The seed coat is fused with fruit wall to form pericarp. The endosperm forms the main bulk of the grain and is the tissue for food storage. It is separated from the embryo by a definite layer known as epithelium.

The outer most layer of the endosperm is the aleurone layer, which unlike the rest of the endosperm, is made up of living cells devoid of galactomannan reserves. This layer secretes alpha-amylase and proteolytic enzymes which hydrolyse reserves of endosperm.

The embryo is very small and lies in a groove at one end of the endosperm. It consist of a shield shaped cotyledon (Scutellum) and a short axis with plumule and radicle protected by root cap.

The plumule as a whole is surrounded by coleptile, a protective sheath, and similarly the radicle including the root cap is surrounded and protected by coleorrhiza. Scutellum supplies growing embryo with food material absorbed from endosperm through epithelium. The initial synthesis of alpha-amylase and certain proteolases also occurs in scutellum.

Seed Growth and Maturation :Wheat and soybean representing monocots and dicots may illustrate the changes in the pattern of accumulation of reserve materials at different stages of seed maturation.

In wheat, the dry weight of the seed increases rapidly in about 35 days after anthesis. The water content of the grain is maximum between 14 and 21 days after anthesis, and then it declines rapidly. The amounts of reducing sugar and sucrose are high between 7 and 14 days and decline rapidly thereafter due to conversion to starch. Since in wheat, starch is the major reserve material of the seed, the pattern of starch accumulation is similar to that of dry matter accumulation. The speed of germination is faster in wheat varieties that begin to lose water early during seed development.

The seed is said to have physiologically matured only when it attains maximum dry weight, germinability and vigour. Normally the seed is harvested at field maturity, a stage when the moisture content is reduced to about 6-10 % in wheat. Field maturity is a crop specific character.

A soybean seed attains maximum dry weight between 48 and 54 days after flowering. Oil accumulation is less during 12-18 days after fertilization; maximum oil accumulates between 24 and 42 days after flowering, after which the rate decreases.

The protein content in the seed is maximum during 12-18 days after fertilization and decreases subsequently. The initial high percentage of protein may be due to the high content of non-protein nitrogen, which decreases with seed.

Seed Coat

It is the outer covering of seed and gives protection. It develops from the 2 integuments of ovule. Outer layer of the seed coat which is smooth and rough is known as the testa and is formed from the outer integument. The inner layer of the seed coat is called the tegmen and is formed from inner integument.

Embryo

It is the mature ovule consisting of an embryonic plant together with a store of food, all surrounded by a protective coat, which gives rise to a plant similar to that of its mother. It is a miniature plant consists of plumule, radicle and cotyledon. The plumule and radical without the cotyledon is known as primary axis.

Radicle

Rudimentary root of a plant compressed in the embryo is the radicle, which forms the primary root of the young seedling. It is enclosed in a protective cover known as coleorhiza.

Plumule

It is the first terminal bud of the plant compressed in the embryo and it gives rise to the first vegetative shoot of the plant. It is enclosed in a protective cover known as coleoptile.

Cotyledon

Cotyledons are the compressed seed leaves. A single cotyledon (Scutellum) is present in monocots while two cotyledons are present in dicots, hence they are named as monocots and dicots, respectively. In dicots they serve as storage tissue and are well developed, while scutellum is a very tiny structure in monocots.

Endosperm

Endosperm develops from the endosperm nuclei which is formed by the two polar nuclei and one sperm nuclei. It stores food for the developing embryo.

Appendages of Seeds

Some seeds will have appentages that are attached to the seed coat. They vary with kind of seed. The appendages sometimes help in dispersal of seeds or in identification of genotypes. Some of the appendages are Awn, Hilum, Caruncle, Aril, Hair and Wings.

Awn: The thorn like projection at tip of the seeds. (eg) Paddy - The bract tip was elongated into the awn.

Hilum: It is the scar mostly white in colour present on the lateral side of the seed. It represents attachment of the seed stalk to placenta of the fruit to mother plant (eg) Pulses.

Micropyle: The point where the integuments meet at the nucellar apex has been referred as micropyle.

Chalaza: At region of integumentary origin and attachment opposite to micropyle is called chalaza.

Rapha: The area between the micropyle and chalaza is the rapha. The rapha may be visible on the seed coat of some species.

Caruncle: It is the white spongy outgrowth of the micropyle seen in some species (eg) Castor, Tapioca.

Aril: It is the coloured flesh mass present on the outside of the seed (eg) Nutmeg.

Hairs: They are the minute thread like appendages present on the surface of the seed (eg) Cotton.

Wings: It is the papery structure attached to the side of the seed coat either to a specific side of the seed coat or to all sides (eg) Moringa.

Quality Seed and Its Importance in Agriculture

What is Seed?

In broad sense, seed is a material which is used for planting or regeneration purpose. However scientifically, Seed is a fertilized matured ovule together covered with seed coat is called seed or it is a propagating material i.e., part of agriculture, sericulture, silviculture and horticultural plants used for sowing or planting purpose. Seed may be defined as "Structurally a true seed is a fertilized matured ovule, consisting of an embryonic plant, a store of food and a protective seed coat, a store of food consists of cotyledons and endosperm"

Definition of Seed

However, from the seed technological point of view seed may be sexually produced matured ovule consisting of an intact embryo, endosperm and or cotyledon with protective covering (seed coat). It also refers to propagating materials of healthy seedlings, tuber, bulbs, rhizome, roots, cuttings, setts, slips, all types of grafts and vegetatively propagating materials used for production purpose. Thus seed is the most vital and crucial input for crop production, one of the ways to increase the productivity without adding appreciably to the extent of land now under cultivation by planting quality seed.

Characteristics/attributes of quality seed;

1. It must be genetically pure:

   Breeder/nucleus: 100 %

   Foundation seeds: 99.5%

   Certified seeds varieties: 98%

   Certified hybrid seeds:95%

   Certified hybrid cotton: 90%

   Certified hybrid castor: 85%

2. It should have required level of physical purity:

   All crops: 98%

   Carrot: 95%

   Ragi: 97%

1. Free form other crop seeds (number/kg): designated inseparable crop seeds are the plants seeds cultivated crops found in the seed fields and whose seeds are so similar to crop seed that is difficult to separate them economically by mechanical means, cause physical admixtures with the crop seeds only when these crop mature approximately at the same time when seed crop matures. Ex. In barley: oats and wheat seeds
2. Free from objectionable weed seeds: these are seeds of weed species which are harmful in one or more of the following ways;
   a. The size and shape of weed seeds are similar to the crop seed and difficult to separate them by mechanical means
   b. Weeds growth habit is determinate type and compete with the crop for all resources

c. Weed plant parts are poisonous or injurious to human and animals

d. Weed platns also serves as alternative hosts for pests and diseases

Ex. Bersem: chicory

Paddy: wild paddy (Oryza sativa var. fatua)

Cucurbits: Wild cucurbits spp.

Lettuce: wild lettuce

Bhendi: wild Abelmoscous spp

Wheat: Convolvulus arvensis (Hirankuri)

3. Free from designated diseases: it refers to the diseases specified for the seed certification which may cause contamination of seed lot

Ex; loose smut of wheat (Ustilago tritici)

Kernal smut of sorghum (Sphacelotheca sorghi)

Grain smut of pearl millet (Tolyposporium penicillariae)

Ergot of bajra (Claviceps microcephala)

Early blight of tomato (Alternaria solani)

Little leaf of brinjal (Datura virus-2) etc.,

4. High germination and vigour

5. Optimum moisture content

Cereals: 10-12 %

Pulses: 7-9%

Oilseeds:6-7%

Vegetables: 5-6%

The importance of quality seeds has been recognised from the time immemorial. The old scripture, Manu Smriti says "Subeejam Sukshetre Jayate Sampadyathe" i.e., Good seed in good soil yields abundantly. Seed quality has been treated as sacred, being an important factor in the improvement of agriculture and agrarian societies. The Rigveda, 2000 BC indicate the importance accorded to seed and the mother earth. 5th century Kautilya Artha Shasthra, surapalas vrikshayurveda mentioned importance of seed and mentioned about seed treatments to ensure good germination. Seed dressing with milk, vidanga, cowdung, honey to protect the seed during germination.

Although the importance of seed was recognised in ancient agriculture, the need for organized seed production was identified only at the beginning of 20th century when Royal commission of Agriculture (1925) recommended spread of improved varieties and seed distribution.

Importance of quality seed:

1. Seed is a vital input in crop production;
   - It is the cheapest input in crop production and key to agriculture progress.
   - Crop status largely depends on the seed materials used for sowing
   - Response of other inputs in crop production depends on seed material used
2. The seed required for raising crop is quite small and its cost is so less compared to other inputs
3. This emphasis the need for increasing the areas under quality seed production
4. It is estimated that good quality seeds to improved varieties can contribute about 20-25% increase in yield.

The advent of modern plant breeding methods and biotechnological advances in seed industry plays a significant role in developing of high yielding varieties and hybrids.

Role of improved seeds:

1. Carrier of new technology
2. Basic tool for a secure food supply
3. The principle means to obtain crop yields in less favourable production area
4. Medium for rapid rehabilitation of agriculture after natural disasters.

Difference between seed and grain:

Seed & Grain

Seed

Grain

It should be a viable

Need not be a viable

It should have maximum genetic & physical purity

Not so

Should satisfy minimum seed certification standards

No such requirements

Treated with pesticide /fungicide to protect seed against storage pests and fungi

Not treated with any chemicals, since used for consumption

Respiration rate and other physiological and biological processes should be kept at low level during storage

No such specifications

Production is technically organized

Not so

It should satisfy all the seed quality attributes

No need

Benefits of using quality seeds

1. They are genetically pure (true to type).
2. The good quality seed has high return per unit area as the genetic potentiality of the crop can be fully exploited.
3. Less infestation of land with weed seed/other crop seeds.
4. Less disease and insect problem.
5. Minimisation of seed/seedling rate i.e., fast and uniform emergence of seedling.
6. They are vigorous, free from pests and disease.
7. They can be adopted themselves for extreme climatic condition and cropping system of the location.
8. The quality seed respond well to the applied fertilizers and nutrients.
9. Uniform in plant population and maturity.
10. Crop raised with quality seed are aesthetically pleasing.
11. Good seed prolongs life of a variety.
12. Yield prediction is very easy.
13. Handling in post-harvest operation will be easy.
14. Preparations of finished products are also better.
15. High produce value and their marketability.

Causes for Varietal Deterioration, Maintenance of Genetic Purity and Principles of Seed Production

General Principles of Seed Production

Production of good quality seed is an exacting task requires high technical skills and heavy financial investments. During seed production strict attention must be given to maintain genetic purity and exploit its potentiality in next generation. In other words, seed production must be carried out under standardized and well organized condition.

Genetic Principles

Causes for Deterioration of Genetic Purity

The genetic purity of a variety or trueness to its type deteriorates due to several factors during the production cycles. Kadam (1942) listed the following important factors responsible for deterioration of varieties.

1. Developmental variations
2. Mechanical mixtures
3. Mutations
4. Natural crossing
5. Minor genetic variations
6. Selected influence of pest and diseases
7. The technique of the plant breeder

    1. *Developmental Variations:* When seed crops are grown under environments with differing soil, fertility, climate photoperiods, or at different elevations for several consecutive generation's developmental variations may set in as differential growth responses. It is therefore, preferred to grow the varieties of crops in the areas of their natural adaptation to minimise developmental shifts.
    2. *Mechanical Mixtures:* Mechanical mixtures, the most important reason for varietal deterioration, often take place at the time of sowing if more than one variety is sown with the same seed drill, through volunteer plants of the same crop in the seed field, or through different varieties grown in adjacent fields. Two varieties growing next to each other field is usually mixed during harvesting and threshing operations. The threshing equipment is often contaminated with seeds of other varieties. Similarly, the gunny bags, seed bins and elevators are also often contaminate, adding to the mechanical mixtures of varieties. Roguing the seed fields critically and using utmost care during seed production and processing are necessary to avoid such mechanical contamination.
    3. *Mutations:* Mutations do not seriously deteriorate varieties. It is often difficult to identify or detect minor mutations occurring naturally. Mutants such as 'fatuoids' in oats or 'rabbit ear' in peas may be removed by roguing from seed plots to purify the seeds.
    4. *Natural Crossing:* Natural crossing can be an important source of varietal deterioration in sexually propagated crops. The extent of contamination depends upon the magnitude of natural cross-pollination. The deterioration sets in due to natural crossing with

undesirable types, diseased plants, or off types. In self-pollinated crops, natural crossing is not a serious source of contamination unless variety is male sterile and is grown in close proximity with other varieties. The natural crossing, however, can be major source of contamination due to natural crossing.

Extent of genetic contamination in seed field due to natural crossing depends up on

1. The breeding system of the species
2. Isolation distance
3. Varietal mass
4. *Pollinating agent:* The isolation of seed crops is the most important factor in avoiding contamination of the cross-pollinated crops. The direction of prevailing winds, the number of insects present and their activity, and mass of varieties are also important considerations is contamination by natural crossing.
5. *Minor Genetic Variations:* Minor genetic variations can occur even in varieties appearing phenotypically uniform and homogenous when released. The variations may lost during later production cycles owing to selective elimination by the nature. The yield trials of lines propagated from plants of breeder's seed to maintain the purity of self-pollinated crop varieties can overcome these minor variations. Due care during the maintenance of nucleus and breeder's seed of cross-pollinated varieties of crop is necessary.
6. *Selected Influence of Pest and Diseases:* New crop varieties often are susceptible to newer races of pests and diseases caused by obligate parasites and thus selectively influence deterioration. The vegetatively propagated stock also can deteriorate quickly if infected by virus, fungi or bacteria. Seed production under strict disease free conditions is therefore essential.
7. *The Techniques of the Plant Breeder:* Serious instabilities may occur in varieties owing to cytogenetic irregularities in the form of improper assessments in the release of new varieties. Premature release of varieties, still segregating for resistance and susceptibility to diseases or other factors can cause significant deterioration of varieties. This failure can be attributed to the variety-testing programme.

In addition to these factors, other heritable variations due to recombination's and polyploidization may also take place in varieties during seed production, which can be avoided by periodical selection during maintenance of the seed stock.

4

Genetic Purity Maintenance

Hartmann and Kester (1968) & Agarwal (1980) described steps to maintain the genetic purity of variety during seed production. The following measures have been suggested to safeguard the genetic purity

1. *Adoption of Crop:* Growing crops only in areas of their adaptation to avoid genetic shifts.
2. *Approved Class of Seeds:* Use of only approved class of seed in seed multiplication and adopt generation system.
3. *Preceding Crop Requirement:* Inspection and approval of seed plots prior to planting
4. *Isolation:* Isolation of seed crops from various sources of contamination by natural crossing or mechanical mixtures.
5. Roguing: Roguing of off types differing in characteristics from those of the seed variety.
6. Field Inspection: Qualified and experienced personnel of seed certification agency should inspect seed crops at all appropriate stages of growth and verify seed lots or purity and quality.
7. GOT: Periodic testing of varieties for genetic purity.

Agronomic Principles

Standardized seed production, besides genetic principles, involves the application of the following agronomic principles to preserve good seed quality and abundant seed yields.

1. Selection of suitable Agro-climatic region
   - The seed crops have to be grown only in areas well adapted to the photoperiodic and temperature conditions prevailing.
   - Region of moderate rainfall and humidity are much more suited to seed production than regions of high rainfall and humidity.
   - In general, regions with extreme summer heat and very cold winters should also be avoided for seed production unless particular crops are especially adapted to grow and product under these conditions.
2. Selection of seed plot
   - Seed production plot should have good texture and fertility.
   - Should be free from volunteer plants weeds and other crop plants.
   - Soil should be free from soil borne diseases and insect pests.
   - The previous crops should not be same crop.
   - The plot should get adequate isolation distance.
3. Isolation of seed crops
   - The seed crop must be isolated from other nearby fields from the same crop or any contaminating crop as per certification standards.
   - Time isolation could also be used in some crops. This is a must to meet the standards for genetic purity of seeds.
   - Isolation of seed production of different varieties is also necessary to avoid mechanical contamination.
4. Preparation of land:
   - The land must be prepared well.
   - Good land preparation helps in improved & uniform germination resulted in good stand establishment.
5. Selection of variety:
   - Variety should be adopted to agro-climatic condition
   - Variety should be high yielder
   - Variety should posses other desirable characters like disease resistance, earliness, grain quality etc
6. Selection of Seed:
   - Seed purity: The seed used for raising a seed crop should be of known purity, appropriate class and invariably obtained from authorized official agency.

7. Seed Treatment:
   — If the seed is not treated already, it should be treated with appropriate fungicides/insecticide prior to sowing.
8. Time of sowing:
   — Should be sown at their normal planting time
9. Seed rate:
   — Lower seed rate than usual to facilitate rogunig operation/seed inspection.
10. Method/depth of sowing:
    — Invariably sown in rows at right depth.
11. Rouging
    — To rouge plants (off types, pollen shedders, diseased plants, etc.,) at the earliest possible but before flowering
    — The number of rouging varies with the crop, purity of the source seed and the stage of the multiplication of the seed crops.
    — Rouging in most of the field crops may be done at - Vegetative, Pre-flowering, Flowering and Maturity stages
12. Supplementary pollination
    — In cross-pollinated crops supplementing the natural pollination is very much essential to increase the seed yield.
    — Provision of honeybees, hand pollination, floral mechanism, and rope pulling are some of the techniques used in crosspollinated crops to increase the seed set and there by seed yield.
13. Weed Control:
    — Compete with seed crop and reduces seed yield and quality
    — Presence of weed seeds at the time of harvest leads to mixing of weed seeds with crops seeds and difficult to separate during seed processing
    — Harbors pests and diseases
14. Disease and Insect management:
    — Reduces seed yield and quality
    — Infection of seeds with spores
15. Nutrition:
    — Important role for proper development of plant and seeds
    — N-Good healthy growth P-Fruiting and seed development K-flowering and seed development
    — Proper nutrition to achieve synchronization

16. Irrigation:
    — Required for translocation of all the nutrients
    — Seed production areas should be dry regions with assured irrigation
17. Harvesting of seed crops
    — Time and method of harvesting are important.
    — Harvest when the seed is fully matured (Moisture content less than 20%)
    — Early or late harvesting affect seed quality.
    — Every effort should be made to avoid chance of mechanical mixing, mechanical injury to seeds during harvesting, threshing, cleaning drying and packing.
18. Drying of seeds
    — In order to preserve seed viability and vigour, it necessary to dry seeds to safe moisture content levels as early as possible without heaping wet seed \ pods.
19. Storage of Raw Seeds. After sun drying, the seed should be filled in clean bags.
    — The bags should be dipped in 2 % Melathion, dried and cleaned before they are filled.
    — Mark the bag with name of variety and other details.
    — The bags should be stacked on wood pallets but not directly on the floors.
    — The height of the stack should not be more than 3 to 4 m for cereals, 2.5 to 3 m for other crops.
    — The godowns should be dry, cool and clean and spray with Melathion and later fumigate as and when necessary.

Seed Multiplication Ratio

It is nothing but the number of seeds to be produced from a single seed when it is sown and harvested. According to expert group on seeds (1989), the seed multiplication ratio for different crops are as follows.

Crop Seed Multiplication Ratio

Wheat 1:20

Paddy 1:80(Variety)

1:100 (Hybrids)

Maize 1:100(Hybrids)

1:80 (Varieties)

Sorghum 1:100

Bajra 1:200

Ragi 1:80

Gram 1:10

lack gram 1:40

Green gram 1:40

Cowpea 1:40

Horse gram 1:40

Moth bean 1:40

Red gram 1:100

Colecrops 1:433

Potato 1:4

Ground nut 1:8

Mustard and rape 1:100

Soy bean 1:16

Sunflower 1:50

Sesame 1:250

Safflower and castor 1:60

Linseed 1:50

Cotton 1:50

Jute 1:100

Mestha 1:40

Sun hemp 1:30

Berseem 1:10

Crops Seed Replacement Rate

Crop SRR (%) Crop SRR (%)

Paddy 29 Cowpea 20

Jowar 23 Fieldgreen 8

Ragi 27 Bengalgram 21

Maize 30 Groundnut 13

Bajra 26 Sesamum 13

Wheat 19 Sunflower 23

Redgram 23 Soyabean 16

Horsegram 7 Safflower 17

Greengram 20 Castor 13

Blackgram 20 Cotton 15

Generation System of Seed Multiplication

Generation system of seed multiplication is nothing but the production of a particular class of seed from specific class of seed up to certified seed stage. The choice of a proper seed multiplication model is the key to further success of a seed programme. This basically depends upon, a. The rate of genetic deterioration b. Seed multiplication ratio and c. Total seed demand (Seed replacement rate) Based on these factors different seed multiplication models may be derived for each crop and the seed multiplication agency should decide how quickly the farmers can be supplied with the seed of newly released varieties, after the nucleus seed stock has been handed over to the concerned agency, so that it may replace the old varieties. In view of the basic factors, the chain of seed multiplication models could be.,

a. THREE - Generation model - Breeder seed - Foundation seed - Certified seed

b. FOUR - Generation model- Breeder seed - Foundation seed (I)- Foundation seed (II) - Certified seed

c. FIVE - Generation model Breeder seed - Foundation seed (I)- Foundation seed (II) -Certified seed (I) - Certified seed (II)

For most of the often cross pollinated and cross pollinated crops 3 & 4 generation models is usually suggested for seed multiplication .e.g. Castor, Red gram, Jute, Green gram, Rape seed, Mustard ,Sesame , Sunflower & most of the vegetable crops. Opel Green Classes of seed

The four generally recognised classes of seeds are: Breeder's seed, foundation seed, registered seed and certified seed.

a) *Nuclus seed:* The initial hand full of seeds obtained from selected individual plants of a particular variety, for the purpose of purifying and maintain that variety, by originating plant breeder.

b) *Breeder's seed:* Progeny of Nucleus seeds, its production is directly controlled by the originating or the sponsoring breeder or institution, providing for the initial and recurring increase of foundation seed.

b) *Foundation seed:* The Progeny of breeders or foundation seed handled to maintain specific genetic purity and identity. This seed is the source of all other certified seed classes.

d) *Certified seed:* The progeny of foundation or certified seed that is handled so as to maintain satisfactorily genetic identity and purity and that has been approved and certified by the certifying agency.

Differences between Certified Seed and Truth Ful Seed

Certified seed

Truthful labelled seed

Certification is voluntary

Truthful labelling is compulsory for notified kind of varieties

Applicable to notified kinds only

Applicable to both notified and released varieties

It should satisfy minimum field and seed standards

Tested for physical purity and germination

Seed certification officer, seed inspectors can take samples for inspection

Seed inspectors alone can take samples for checking.

Methods for Hybrid Seed Production

Hybrid is produced by crossing between two genetically dissimilar parents. Pollen from male parent (Pollen parent) will pollinate, fertilize and set seeds in female (seed parent) to produce F1 hybrid seeds. For production of a hybrid CROSSING between two parents is important, the crossing process will results in heterosis. In self pollinated cross it is difficult to cross but in cross pollinated crops it is easier.

In nature to create genetic variability and for its wider adaptation in different environmental conditions, flowering plants has adopted many mechanisms for cross pollination. Cross-pollination results in genetic heterogeneity and show wider adaptations. Flowering plants have evolved a number of devises to encourage cross-pollination. Those mechanisms are;

1. *Dicliny:* Flowers are unisexual. In monoecious plants male and female flowers are borne on the same plant eg., cucurbits, maize, castor and coconut. In dioecious plants male flowers are borne on different plants eg., papaya, cannabis, mulberry.
2. *Dichogamy:* Time of anther dehiscence and stigma receptivity are different forcing them for cross-pollination. The time gap between the two may vary from one day to many days. In protoandry anthers dehisce earlier than the stigma receptivity eg; maize, sunflower. In protogyny stigma become recetive earlier than the anther dehisce eg., Pearl millet mirabilis.

3. *Self-incompatibility:* self fertilization in avoided by recognising the self pollen by the stigma. Eg., Brassica, Petunia, Lilium.
4. *Herkogamy:* there is spatial separation of the anthers and stigma. Their relative position is such that self fertilization cannot occur. The stigma projects beyond the anthers and therefore pollen cannot land on stigma. Eg., Lucerne stigma is covered with a waxy film. The stigma does not become receptive until this waxy membrane is broken by visit of honeybees resulting in cross-pollination.
5. *Male sterility:* Absence or atropy or mis or malformed of male sex organ (functional pollen) in normal bisexual flower. Male sterility is of three types: genetic male sterility, cytoplasm sterility and cytoplasmic- genetic male sterility.
6. A combination of two or more of the above mechanisms may occur in some species. This improves the efficiency of the system in promoting cross-pollination

Requisites of hybrid seed production:

1. Breeders responsibilities:
   (a) Develop inbred lines
   (b) Identification of specific parental lines
   (c) Develop system for pollen control
2. Major problems for breeders & producers
   (a) Maintenance of parental lines
   (b) Separation of male and female reproductive organs
   (c) Pollination

Characteristics of parental lines:

Female Parent Male Parent

High seed yield Good pollen production

Good seed characteristics Long shedding period

Male sterility Plant height

Lodging resistant Fertility restoration

Basic procedures for hybrid seed production

1. Development and identification for parental lines
2. Multiplication of parental lines
3. crossing between parental lines and production of F1

Commercial hybrid seed production demands crossing technique which is easy and also economic to maintain parental lines. Only few crossing mechanisms have been adopted for commercial hybrid seed production they are;

1. Hand emasculation and pollination
2. Self-incompatibility
3. Dicliny : monoecious and dioecious
4. Male sterility

These techniques are specific to crop floral biology and flowering behaviour. These techniques have their own advantages and disadvantages. Based on the crop behaviour and crossing technique have been adapted for production of hybrid seeds commercially.

Among different techniques self-incompatibility and sex expression have significance particularly in vegetable and flower hybrid seed production.

1. *Hand emasculation and pollination:* Hybrid seeds are produced manually by modifying the plant structure by removal of male organ from female plant before anthesis. This system is possible only when the male and female parts of a single flower or plants are separate. This is being adopted in bisexual perfect flowers where the androecium is removal with case. By removing the anther column / or male part from female line, the sterility of female line is created and is dusted with the pollen of desired male parent.
2. *Self Incompatibility:* Self-incompatibility is a mechanism which avoids self fertilization through recognition of self pollen in or on stigma on the female pistil. But when pollen from other plant carried by wind or insects are accepted and sets seeds. Self-incompatibility will prevents self pollination (inbreeding) and promotes crosspollination (out breeding) and creates genetic variability. SI are seen in hermaphrodite and homomorphic flowers. Self-incompatibility is a widespread mechanism in flowering plants that prevents inbreeding and promotes outcrossing. The self-incompatibility response is genetically controlled by one or more multi-allelic loci, and relies on a series of complex cellular interactions between the self-incompatible pollen and pistil.

Types of Self-Incompatibility (SI)

I. Heteromorphic Self-Incompatibility: In this system flowers are of different morphology of the reproductive parts. The morphological differences can be seen visibly in flowers this will coincide with crossibility. The characters affecting this type of SI are style length, filament length, pollen size, exine sculpturing. The presence or absence of other SI mechanisms will not affect cross pollination in heteromorphic SI.

a. *Heteromorphic distyle:* eg., Primula: The flowers of primula have style at two different heights. Has two types of flowers

1. Thrum flower: short style and long anther
2. Pin flower: has short anthers and long style

   But these two are cross compatible.

   Pin x Pin (ss x ss) : Incompatible mating

   Pin x Thrum (ss x Ss) : Compatible mating

   Thrum x Pin (Ss x ss) : Compatible mating

   Thrum x Thrum (Ss x Ss) : Incompatible mating

b. Heteromorphic Tristyle: eg., Lythrum

   Three types of flowers;

   1. Short style and stamens are mid and long
   2. Mid style and stamen are short and long
   3. Long style and stamen are short and mid

When homomorphic flowers are crossed it results in incompatible mating and when hetromorphic flowers are crossed it results in compatible mating.

II. Homomorphic Flowers: Flowers morphological are same, so mating types cannot be recognised by morphological features. This types of self-incompatibility is controlled by 'S' alleles. For crossing parental 'S' alleles should be different, then only fertilization takes places and seed sets. Two types of self-incompatibility;

1. Sporophytic self-incompatibility
2. Gametophytic self-incompatibility

Sporophytic Self-Incompatibility (SSI)

SSI is less common or rare when compared to GSI. The rejection of pollen is controlled by S loci which is dominant. The dominance relationship are like S1> S2>S3>....... This type of self-incompatibility is controlled by diploid genotypes of the sporophyte (pollen). Pollen will not germinate on the stigma of the flower that contains either of the two alleles in the pollen so, rejected.

Gametophytic Self-Incompatibility (GSI)

GSSI is more common when compared to SSI. The rejection and acceptance of pollen is controlled by 'S' loci. Unlike SSI, incompatibility is controlled by haploid genotype of pollen itself.

S1 pollen can germinate on pistil of S1S2 but due to common S1 allele the pollen tube growth seizes. Similarly in S2 the pollen tube growth seizes. When S3 pollen come in contact with pistil of S1S2 there will be normal growth of pollen tube and fertilization takes place, this is called as partial compatibility.

Where as, S3 and S4 pollen can pierce in to the style (S1 S2) and cause fertilization, called as complete compatibility.

In single gene system there are three types of mating systems they are;

S1S2 X S1S2 : 0% compatibility

S1S2 X S2S3 : 50% compatibility

S1S2 X S3S4 : 100% compatibility

Pollen rejection and acceptance of cross pollen is a mechanism of complex interactions.

There are three main interactions;

1. Pollen –stigma interaction seen in SSI
2. Pollen- style interaction seen in GSI
3. Pollen tube- ovule interaction seen in GSI
4. Pollen tube- ovule interaction is very rare and seen in some cases of cocoa, pollen tube reaches ovule and effect fertilization but incompatible combination will degenerate the embryo at early stage development only resulting in no seed set.

    1. *Pollen-stigma interaction:* This type of interaction is seen in SSI. It includes dry stigma. The stigma has a hydrated layer of proteins know as pellicle. This pellicle is involved in incompatible reaction. Within a few minutes of pollen reaching stigmatic surface, the pollen releases an exine exudates which is glycoprotein. Due to 'S' allele proteins interactions induces immediate callose formation in the papillae, this is in direct contact with pollen. Also this callose is deposited on the protruding pollen tube preventing further germination of the pollen. Here stigma is the site of incompatibility reaction. If once pollen cross this stigmatic barrier, there is no further rejection of pollen.

       In case of wet stigma seen in GSI, when pollen sits on the stigma, pollen coat stimulate an unusual form of stigmatic reaction focused beneath the areas of coating. This results in the deposition of fibrogranular electron opaque layer by the extracellular vesicles in the outer layer of the papillae wall. This results in blockage of water passage or loss of osmotic competence by pollen resulting in drying of pollen.

       Pollen tube - style interaction: This is common in gametophytic self-incompatibility. The pollen can germinate and pollen tube growth will pierce the stigma. But the rate of the pollen tube growth is very slow when compared to compatible pollen. This slow growing pollen tube will stops due to exhaustion of reserve materials and deposition of 'S' allele polysaccharide at the tip of the pollen tube which blocks the growth of tube. This deposition is limited to the tip and thus will not affect the other compatible pollen tube growth.

2. Three-way hybrid: Two SI lines are crossed to get an heterozygote and again crossed with self-compatible parent which has suppressors locus and give F1 hybrid which is self-compatible
3. Double cross hybrid
4. Triline hybrid: Hybrid seed production is done by crossing of two inbreed lines. Have to maintenance these inbreed lines which are self-incompatible. For success of self pollination need to eliminate SI. There are many methods to over come SI
   - Bud pollination (Tatebe, 1939)
   - Washing stigma surface/ pollen grains with organic solvents (Roggen 1974)
   - Mechanical and electric methods (Roggen & Van Dijk, 1972; 1973)
   - CO_2 Treatment (Nakanishi et al., 1969)
   - High temperature treatment (Visser, 1977)

Modification of sex: Hermophrodite flowers has both male and female reproductive organ in a single flower.

1. Complete flower; flowers containing all the four whorls viz., sepals, petals, androecium and gyneocium.
2. Incomplete flower: flowers missing any of the one whorl.
3. Imperfect Flowers- Sexual distinctness: Monoecious (Maize) vs. Dioecious (Hollies, Poplars)
4. Perfect Flowers- Gamete Maturation Time (Dicogamy): Production of this unisexual/ imperfect flowers will naturally leads;
   - Outcrossing avoids the deleterious effects of inbreeding depression
   - Promotes heterozygosity, genetic variability, and genetic exchange,
   - Advantageous to the long-term survival and adaptation of a species.

Monoecious: Flowers are unisexual and are present at different position on the same plant. Eg. cucumber. Terminal flowers are male flower. In the middle of the plant is female favouring crosspollination. Dioecious: male flowers and female flowers are in different plant. So called as male plant and female plant.

Sex Modification through Hormones and Chemicals

Sex expression in dioecious and monoecious plants is genetically determined and can be modified to a considerable extent by environmental and introduced factors such as mineral nutrition, photoperiod, temperature, phytohormones. Amongst these, phytohormones have been found to be most effective agents for sex modification and their role in regulation of sex expression in flowering plants has been documented. The morphological differences in various sex types and their specific metabolic characteristics result from the possession of specific patterns of proteins, enzymes and other molecules. Modification of sex expression in cucurbits has been induced both by changing the environmental conditions and by applying treatments with growth regulators. Auxin treatments increase the female sex tendency while gibberellins cause a shift towards maleness.

Hormones & Chemicals inducing Femaleness: Auxins- NAA, Etherl, Ethephon, Cytokinis- BA, Brassinosteriods etc. Hormones & Chemicals inducing Maleness: GA3, AgNO3, ABA Thio porpinic acid, Pthalimide, Paclobutrazol etc.

In cucumber AgNO3 found to be potent inhibitors of ethylene action leading to femaleness. It should be sprayed when first true leaf is fully expanded. Gibberlic acid spray will leads to excessive elongation and weakening of plants and there will be increased number of mall formed male flowers with less pollen. In gynoecious cucumber there will be increased number of male nodes when sprayed with silvernitrate and gibberlic acid, which made possible for multiplication of gynoecious in hybrid seed production.

Environmental Sex Modification

Environment has greater influence on the sex modification. But due to introduction of photosensitive varieties or hybrids in modern era of agriculture it has gained less importance. However in seed production it has its influence on sex expression. In cucumber, high temperature and long day length (> 14 hours) favours male flowers. High temperature will extends the flowering of female flowers. As the temperature increases form 19°C to 23°C the node for first female flowers has also increased from 9.6 to 16.5 number. This clearly indicates that high temperature favours male and delays female flowering.

Male sex expression of several plant species is favoured by high temperatures and female sex expression by low temperatures. Male sterile mutant of tomato developing male sterile flowers at a minimum temperature of 30°C and normal flowers at lower temperatures. In Brussels sprouts of low temperature effect on the development of the androecium. In onions a slight production of viable pollen by normally male sterile plants above 20°C.

Male Sterility

Hybrid production requires a female plant in which no viable male gametes are borne. Emasculation is done to make a plant devoid of pollen so that it is made female. Another simple way to establish a female line for hybrid seed production is to identify or create a line that is unable to produce viable pollen. This male sterile line is therefore unable to self-pollinate and seed formation is dependent upon pollen from the male line.

In hermaphrodite flowers pollens are non-functional or inactive or sterile while, female gametes functions normally. It is the inability of plant to produce or to release functional pollen as a result of failure of formation or development of functional stamens, microspores or gametes. Male sterility can be either genetic or cytoplasmic or cytoplasmic-genetic. This prevents autogamy and permits crosspollination. Promotes heterozygosity. Sterility is due to nuclear genes or Cytoplasmic gene or both.

In hybrid seed production process female is a male sterile line crossed with male fertility restorer line to get F¬1 hybrid.

Cytoplasmic Male Sterility

Cytoplasmic male sterility, as the name indicates, is under extra nuclear genetic control mainly mitochondrial genome. They show non-Mendelian inheritance and are under the regulation of cytoplasmic factors. In this type, male sterility is inherited maternally. In general there are two types of cytoplasm: N (normal) and the aberrant S (sterile) cytoplasms. These types exhibit reciprocal differences. Cytoplasmic male sterility (CMS) is caused by the extra nuclear genome (mitochondria or chloroplast) and shows maternal inheritance. Manifestation of male sterility in CMS may be either entirely controlled by cytoplasmic factors or by the interaction between cytoplasmic and nuclear factors.

Stamen (anther and filament) and pollen grains are affected. It is divided into:

a. *Autoplasmic:* CMS has arisen within a species as a result of spontaneous mutational changes in the cytoplasm, most likely in the mitochondrial genome

b. *Alloplasmic:* CMS has arisen from intergeneric, interspecific or occasionally intraspecific crosses and where the male sterility can be interpreted as being due to incompatibility or poor co-operation between nuclear genome of one species and the organellar genome another CMS can be a result of interspecific protoplast fusion. Cytoplasmic male sterility is used in hybrid seed production. In this case, the sterility is transmitted only through the female and all progeny will be sterile.

This is not a problem for crops such as onions or carrots where the commodity harvested from the F1 generation is produced during vegetative growth. These CMS lines must be maintained by repeated crossing to a sister line (known as the maintainer line) that is genetically identical except that it possesses normal cytoplasm and is therefore male fertile.

Disadvantages:

1. insufficient or unstable male sterile
2. Difficulties in restoration system
3. Difficulties with seed production

Cytoplasmic-Genetic Male Sterility

Male sterility is controlled by an extranuclear genome and often nuclear genes can have the capability to restore fertility. When nuclear restorations of fertility genes ("Rf") are available for CMS system in any crop, it is cytoplasmic-genetic male sterility; the sterility is manifested by the influence of both nuclear (Mendelian inheritance) and cytoplasmic (maternally inherited) genes. There are also restorers of fertility (Rf) genes, which are distinct from genetic male sterility genes.

The Rf genes do not have any expression of their own unless the sterile cytoplasm is present. Rf genes are required to restore fertility in S cytoplasm which causes sterility. Thus N cytoplasm is always fertile and S cytoplasm with genotype Rf- produces fertile; while S cytoplasm with rfrf produces only male steriles. Another feature of these systems is that Rf mutations (i.e., mutations to rf or no fertility restoration) are frequent, so N cytoplasm with Rfrf is best for stable fertility.

Cytoplasmic-genetic male sterility systems are widely exploited in crop plants for hybrid breeding due to the convenience to control the sterility expression by manipulating the gene–cytoplasm combinations in any selected genotype. Incorporation of these systems for male sterility evades the need for emasculation in cross-pollinated species, thus encouraging cross breeding producing only hybrid seeds under natural conditions.

In cytoplasmic-genetic male sterility restoration of fertility is done using restorer lines carrying nuclear restorer genes in crops. The male sterile line is maintained by crossing with a maintainer line which has the same genome as that of the MS line but carrying normal fertile cytoplasm.

Genetic Male Sterility: Male sterility is controlled by mutations in nuclear genes in the single recessive genes affect stamen and pollen development, but it can be regulated also by dominant genes. MS alleles are generally recessive. A male sterile line is maintained by crossing with heterozygous male fertile line.

Male sterile plants of monoecious or hermaprodite crops are potentially useful in hybrid program because they eliminate the labour intensive process of flower emasculation.

Constraint of the Use of Genetic Male Sterility: The maintenance of the male sterile line. Normally, a GMS line (A-line) is maintained by backcrossing with the heterozygote B-lines (Maintainer lines), but the progeny produced are 50% fertile and 50% male sterile

Solution:

1. Identify marker genes that are closely linked to ms genes and affect some vegetative characters
2. Use of environmental and chemical methods that can lead to production of 100% male-sterile seed

Modification of Sex

Hand emasculation and pollination of individual flowers is the most expensive and hence to overcome this mechanism sex expression is adopted. Hermophrodite flowers has both male and female reproductive organ in a single flower.

1. Complete flower; flowers containing all the four whorls viz., sepals, petals, androecium and gyneocium.
2. Incomplete flower: flowers missing any of the one whorl.
3. Imperfect Flowers- Sexual distinctness: Monoecious (Maize) vs. Dioecious (Hollies, Poplars)
4. Perfect Flowers- Gamete Maturation Time (Dicogamy): Production of this unisexual/ imperfect flowers will naturally leads;
   — Outcrossing avoids the deleterious effects of inbreeding depression
   — Promotes heterozygosity, genetic variability, and genetic exchange,
   — Advantageous to the long-term survival and adaptation of a species.

Monoecious: Flowers are unisexual and are present at different position on the same plant. Eg. cucumber. Terminal flowers are male flower. In the middle of the plant is female favouring crosspollination.

Dioecious: Male flowers and female flowers are in different plant. So called as male plant and female plant. Eg: papaya.

Vegetables

Seed Production on Cole Crops

Cole crops are temperate vegetables also known as European vegetables requiring temperate climate especially during a specific stage of their growth for successful seed production. During this period that these vegetables must have the vernalization requirement a pre condition necessary for breaking the dormancy of the plant thus stimulating the conversion of the vegetative phase into the reproductive phase i.e., induction of bolting and flowering.

All Cole crops are hardy and thrive best under condition of cool climate. They are grown in the plain during winter season and can be cultivated throughout the year in the hill regions of the country.

Origin: All crops of this group have originated from single progenitor wild cliff cabbage known as Cole worts (Brassica oleraceae var sylvistris.) as a result of mutation. All the Cole crops belong to the same genus brassica and most of them to the same species.

Cole crops are Mediterranean in origin and are spread all over European and have been cultivated continuously since 2500 BC. Curly Kale, Cabbage, Knol kohl were the first known cultivated types and other were cultivated in later centuries.

Cauliflower became widely known in 18th century and Brussels sprouts in the 19th century. In the 30's of the 20th century, another form 'Sprouting Broccoli' in USA. Now Cole crops are grown in every corner of the world.

Different Types of Cole Crops

Sl. Crops Botanical Name Chromosome number (n) Plant parts used

1. Cabbage Brassica oleraceae (L) var. capitata 9 Heads
2. Cauliflower Brassica oleraceae (L) var. botrytis.L. 9 Curd
3. Knol Kohl Brassica oleraceae (L) var. gongylodes. L. 9 Swollen stem
4. Brussels sprouts Brassica oleraceae (L) var. gemnifera 9 Sprouts (Swollen axillary buds)
5. Sprouting broccoli Brassica oleraceae (L) var. italica 9 Flower heads
6. Kale Brassica oleraceae (L) var. acephala 9 Top leaves

Constraints for seed production:

- Problems of satisfactory isolation due to cross-pollination by insects.
- Crops have to be carried over in to the second season
- Plant attains morphological shape and size during the additional growing period, which is not known to majority of the seed growers.

Major seed production areas in the world.

1. *North western Europe:* Eastern and South Eastern, UK, Holland, Western Denmark
2. *Central Europe:* Southern Sweden, Eastern Denmark, North Germany, Switzerland

3. *Mediterranean area:* Southern France, Southern Italy
4. *North America and Canada:* Western USA (Pacific North West), Canada (British Columbia)
5. California.

Seed production areas in India:

1. Srinagar valley (J&K)
2. Upper Kullu valley (Himachal Pradesh)
3. Lahaus valley (Himachal Pradesh)
4. Kalpa valley, Kinnaur (Himachal Pradesh)
5. Saproon valley, Solan (Himachal Pradesh)
6. Kumaon hills (Uttar Pradesh)
7. Kalimpong – Darjeeling hills (West Bengal)
8. Nilgiris (South India)

In 1942-43 for the first time imperial Government encouraged the seed production of European type of vegetable at Quetta in Baluchistan. At about the same time initial trials on seed production were also initiated in Kashmir, Katrain (Kullu valley) (Verma and Sharma, 1999).

Inflorescence: After leafy growth ceases, as for example the completed growth of the head of the cabbage, or the sprouts of Brussels sprouts, the flowering stem elongates. It is characterized by numerous branches (mostly from a main stem), small leaves, and numerous bright yellow or occasionally white flowers. The flowers of all Cruciferae have four petals, l/2 to 1 inch long, that appear to form a cross, hence the name Cruciferae (cross bearing).

The flowers are typical of crucifers' family having 4 petals, 6 stamens, of which 2 are short and 2 carpels with superior ovary. The flowers are attached with the stalk with a short pedicel. The anthesis of flowers depends on temperature and flowers open mainly during hotter part of the day. The flower opens during the morning, the anthers a few hours later, so the flower is slightly protogynous. The flowers are highly attractive to pollinating insects for both nectar and pollen. When the seed-producing acreage is large, beekeepers nearby frequently harvest a crop of excellent honey. The stigma of Brassica spp. is receptive even 5 days before and 4 days after anthesis. The period from pollination to fertilization generally takes 24-28 hours, depending on temperature. Higher temperature during daytime is more harm full for fertilization and this causes pollen sterility. The pod maturity or harvest of pods may require 50-90 days from the date of flowering. The seeds are small, globular, smooth and dark brown in colour. The Cole crops are highly cross-pollinated the percentage of cross-pollination varies with the type of crops. The Cole crops are highly cross-pollinated as characterized by both self-incompatibility and protogyny. The self-incompatibility is of sporophytic type, and it is using for the hybrid seed production program, self-pollination of Brassica resulted in decreased yields in subsequent generations.

Percent of cross pollination in Cole crops.

Crop Cross percentage

Cabbage 73%

Cauliflower 70%

Knol kohl 91%

Sprouting broccoli 95%

Brussels sprouts 72%

Kale 83%

Flowering season and Vernalization temperature

Sl.No. Crop Flowering season Vernalization

1. Cabbage March - May 4.4-10°C
2. Cauliflower February- April 10-16°C
3. Knol kohl March - May 4-10°C
4. Sprouting broccoli February- April 10-15°C
5. Brussels sprouts April-May 2.2-10°C
6. Kale March- April 4-8°C

The blossom forms a siliqua, incorrectly but commonly called a pod.

Vernalization

Vernalization of Brassica oleraceae is performed on mature vegetative plants, which are uprooted in autumn from the fields, potted and over wintered in a greenhouse at temperatures between 5-10°C. The cabbages are decapitated (±3 cm of head remains). Other Brassicas are defoliated, leaving ±6 leaves, to prevent rotting. Early cabbages are decapitated in the field, and the new growing shoots are cut and rooted. The cuttings are placed in a cooling compartment for 1 day, to desiccate the cutting surface. No leaves are removed, to avoid the creation of fresh tissue wounds. The cuttings are planted in trays with soil and covered with cheesecloth. After roots are formed, the cuttings are potted. The plants from these rooted cuttings over winter.

Isolation

In brassica there are two distinct groups which do not intercross with each other, Brassica oleraceae with all its botanical varieties and wild allies in one group and B. rape, B.junceae, B, chinensis, etc., from another group.

Cauliflower

India is the largest producer of Cauliflower in the world (FAO, 2001). With the development of tropical types in Cauliflower in addition to the temperate type, it has now become possible to grow this vegetable almost throughout the year particularly in the northern and central part of India. Cauliflower is thought to have been domesticated in the Mediterranean region since the greatest range of variability in type of brassica oleraceae. It became more commonly cultivated in the beginning of the 18th century. Indian or tropical cauliflower, which are early maturity and wider adaptability to hot and humid weather conditions. Temperate types are known as 'Snow ball' or late cauliflower. Cauliflower is a monogenic species whose genomic constitute is C and n=9.

Varieties are very specific to the season when the curds are formed; they are two major groups of cauliflower;

1. Early or Indian cauliflower: are annual types
2. Late or Snow ball types: biennial types. They require low temperature for transition to generative phase i.e., curd initiation stage. Their behaviour to curding is like cabbage for transition to reproductive stage.

Early group is further divided into three maturity groups on the basis of period of their curd availability. I, II, III require high temperature for growth and curd formation where as late group produces curds at low temperature.

Climate: Requires cool, moist climate for seed production. The optimum monthly average temperature is 15 to 20°C, the early varieties, however, require high temperature and longer day lengths.

It is less tolerant to extreme high or low temperatures, or strong winds. It is also susceptible to cold injury after the curds have appeared. Excessive rains and snowfall, after curd formation, cause rotting in curds. Periods of low temperature are not essential, but cool conditions are required. Therefore, these conditions must be given due consideration in selecting suitable areas for seed production. In India, the seed production of early and mid-season varieties can be done in the plains. However, the seed of late varieties can only be produced in temperate regions of the country. Lately, Himachal Pradesh has emerged as the major producer of quality cauliflower seed of late varieties.

Land Requirements:

1. Land to be used for seed production shall be free of volunteer plants.
2. The soil of selected field should be deep, fertile, well-supplied with organic matter with a pH value 5.5
3. The field should be well drained.

Isolation Requirements: Cauliflower is mainly cross-pollinated. Pollination is mainly done by bees. The seed fields must be separated from fields of other varieties, fields of the same variety not conforming to varietal purity requirements of certification, and from all their kinds of Cole crops, at least by 1600 metres for foundation seed class, and by 1000 metres for certified seed class.

Method of Varietal Seed Production:

1. In situ method (seed to seed method)
2. Transplanting method (head to seed method)

For seed production, seed to seed method is recommended since the head to seed method in India has not been very successful. In seed to seed method (in situ method) the crop is allowed to over-winter and produce seed in the original position, where they are first planted in the seedlings stage. Main season and late varieties (seed production in hills)

Time of Sowing and Transplanting: In the hills, the sowing time of cauliflower should be so adjusted that the plants put up the maximum leafy growth by the fifteenth of December, when the temperature goes down and the plants become almost dormant. The last week of August is the optimum sowing time for the crop. The seed is sown in a nursery. Transplanting of seedlings should be completed by the end of September. The mean temperature of 6.5 to 11 C during February to March is very conducive to curd formation, which is completed by the first fortnight of March.

Early sowing in June to July result in curd formation during October to November. The curds, being very susceptible to cold injury, rot during winter and hence fail to flower the following summer. If sown late, the crop starts curd formation late in the spring and consequently flowering is delayed. It starts when the temperature is high and humidity is low, with the results that pollination and setting of seed is not normal.

Method of Sowing in Nursery: Seeds may be sown on raised nursery beds (15 to 20 cm high from the ground), in rows 5 cm apart. Cover the seeds with fine leaf mould and water with a sprinkler. Twenty-five nursery beds of 2 to 2.5 metres long and 1 to 1.25 metres wide will raise enough seedlings to plant one hectare. A spray of four to five handfuls of ammonium sulphate or C.A.N. dissolved in 30 to 35 liters of water at 10 to 15 days after germination will be helpful in producing healthy and vigorous seedlings. Wash out the fertilizer immediately by spraying simple water. Thin sowing should be done to avoid "damping-off" disease.

Source of Seed: Obtain breeder's /foundation seed from source approved by a seed certification agency.

Seed Rate: Mid season and late varieties – 375 to 400 g /ha.

Preparation of Land for Transplanting: The field should be prepared to fine tilth by deep ploughing, three to four harrowing followed by levelling.

Fertilization: The cauliflower seed crop required heavy manuring as it removes large quantities of major nutrients from the soil. For best results apply 50 to 60 tones of farmyard manure at the time of land preparation. Normally 25 to 30 tones of farmyard manure is applied per hectare due to limited availability of farmyard manure in hills. Apply 200 to 300 kg superphosphate and 100 kg potassium sulphate sufficiently before sulphate, or calcium ammonium nitrate, during growing period (one application during October to November, another in February to March is essential). Still higher doses of nitrogen may be applied it deemed necessary. Cultural practices determine greatly the seed yield and among the various inputs, nitrogen and phosphorus application have been reported to exert a great influence on the seed yield of Cauliflower.

A cauliflower crop often shows boron and molybdenum deficiency symptoms when grown either on an alkaline or highly acidic soil, or two sprays with 0.3 per cent borax applied on the seedlings may correct the boron deficiency. Molybdenum deficiency symptoms occur in highly acidic soils and can be corrected by liming, or application of about 1 to 1.5 kg per hectare of sodium molybdate. Spraying of 150 ppm Ethrel at the time of emergence of flowering stalks increase seed yield.

Transplanting: Transplant the seedlings when 12 to 15 cm long, preferably at evening time, and irrigate immediately afterwards.

Spacing:

Row to row 60 to 90 cm and

Plant to plant 45 to 60 cm.

Irrigation: Irrigate the field according to the soil requirements and climatic conditions. A crop after transplanting may need irrigation twice a week and later once a week. At later stages, irrigation may be given if there is a long gap between rains. Adequate moisture supply during flowering and seed maturation are necessary to obtain high yields.

Intercultural: Frequent shallow cultivation should be given to the soil to kill weeds and provide soil mulch. Earthing-up of plants four to five weeks after transplanting is highly desirable.

The seed production is dependent upon curd formation and its bolting. Bolting further depends on the curd type i.e., degree of compactness of the curd. The loose curd bolts easily while compact curds takes more time to bolt.

To facilitate bolting, different curd- cutting methods like scooping, half curd cutting and curd pruning are recommended. These practices have impact on branching, seed yield and seed quality.

Scooping (Single Scoop): Approximately half of the curd was removed from the centre with the help of a sharp knife.

Curd Pruning: The outermost curdlets were pruned 5cms, away from the centre.

Half- curd removal: The curd was cut vertically into two parts from the middle and one of them was removed leaving half portion of the curd intact for seed production.

Roguing: Selection of curds is done when the curds are well developed. Off-type plants, and those forming poor curds, should be removed at this stage. Subsequent roguings for off-types, and diseased plants affected by blackleg, black rot, leaf spot and phyllody should be done from time to time as required.

The developmental process in Cauliflower is regulated by temperature. Low temperature stimulates curd induction at the end of a juvenile phase and further development of the curd, stalk, flower and seed proceed sequentially at rates determined by prevailing temperatures. Seed production in Cauliflower is difficult because of its seasonality, which restrict field production both in the tropics and temperate regions. In tropics, seeds are produced in the cool season or at high altitude where temperatures are relatively low.

Harvesting and Threshing: Harvesting is done when pods brown. Too ripe pods dehisce. Seed should not crush or split when rubbed between the hands. The harvesting maybe done in two lots. Generally the early plants are harvested first, when about 60 to 70 percent of the pods turns brown and the rest of the crop changes to yellowish-brown. After harvesting it is piled up for curing. After four to five days it is turned upside down and allowed to cure for another four to five days in the same way. It is then threshed with sticks and sifted with hand sifters. After thorough drying of seed in the sun (seven percent moisture content) it is cleaned and stored.

Seed Yield: Average seed yield varies from 250 to 400 kg per hectare

Cabbage

Cabbage is one of the important vegetables grown on a large scale in India, which is formed by the development of densely overlapped leaves around the growing point. Cauliflower is mostly grown in North India whereas cabbage is popular in the south and South Eastern parts on India. Cabbage has been under cultivation since 2500 BC to 2000 BC. It is a biennial in nature having

2 specific periods of growth namely vegetative and reproductive phases to complete its life cycle. In the first season crop remains in the vegetative phase characterized by the formation of heads where as the reproductive phase is completed in the second season only after getting the necessary stimuli of low temperature of 4.40C-100C for about 60 days to break the dormancy of the heads. Cabbage is commonly used fresh as salad, Cole slaw, boiled vegetable cooked in curries and processed.

Floral Biology of Cabbage: Cabbage flowers are borne in terminal raceme, flowers are yellow in colour. They are hypogynous having slender pedicels. The flowers are perfect, regular with 4 sepals, 4 petals, 6 stamens, ovary two celled with several ovules per cell. Cabbage starts flowering in the month of February- march or earlier. The stigma remains receptive for about 5 days before and 4 days after anthesis. Cabbage fruits is siliqua, having a thin partition dividing it lengthwise, along with it dehisce when fully mature and dry. Pod contains about 12-20 seeds.

Cultivated forms of Cabbage

White cabbage: Brassica oleraceae Var.

Capitata L.Falba Red cabbage: Brassica oleraceae Var.

Capitata L.frubra Savoy cabbage: Brassica oleraceae Var. sabauda L.

Based on maturity time. Earlyseason: it takes 60-70 days for maturity.

Golden acre, Pride of India, Copenhagen market and Early drumhead.

Mid- season: cabbage comes to harvest at 80-90 days.

All head early, Wisconsin, All green september.

Late season: Require long winter season comes to harvest at 90-120 days

Pusa drum head, Indian eclipse, Danish ball head, Late flat and Dutch sure head.

Based on shape of heads.

Round head or Ball Head type (early): Golden Acre, Pride of India, Copenhagen Market, Mimmothi, Rock red and express.

Flat head or Drumhead type (Late): Pusa drumhead.

Conical head (early to mid season): Jersy wake field

Savoy type (late varieties): Chieftain

Climate: Cabbage thrives in a relatively cool, moist climate with moderate to heavy rainfall, well distributed during the growing season. It can withstand frost in the head stage, but otherwise freezing temperatures are destructive. It requires a dormant period of cool temperature to bolt and initiate seed stalks and flowers. Cool temperatures, however, are effective only after stem diametre is one cm, at least. In temperate climates, this occurs during the winter after the first seasons growth. Flowering and seed production follow in the second year. Headed plants form seed stalks when exposed to mean temperature of about 5C for six to eight weeks. As little as two weeks of such temperature suffice with immature plants. In India, seed production of cabbage is possible only in hill areas.

Methods of Seed Production: Being a biennial, the cabbage requires two seasons to produce seed. In the first season the heads are produce, and in the following season seed production follows. The seed crop can be left in situ or transplanted during autumn. In situ method is usually followed for certified seed production and the latter for nucleus production.

In the in situ method, the crop is allowed to over-winter and produce seed in their original position, that is, where they are first planted in the seedling stage. In the transplanting method, the mature plants are uprooted. After removing whorls the plants are immediately reset in a well-prepared new field, in such a way that the whole stem below the head goes underground with the head resting just above the surface. Cabbage is one of the important vegetables grown on a larg. Three methods have been devised to produce seed of cabbage.

Stump Method: In this method, when the crop in the first season is fully mature, the heads are examined for trueness to type. The plants with off-type heads are removed. Then heads are cut just below the base by means of a sharp knife, keeping the stem with outer whorl of leaves intact. The beheaded portion of the plant is called "stump". The heads are marketed and the stumps either are left in situ, or replanted in the second season i.e., during autumn. The following spring, after the dormancy is broken, the buds sprout from the axils of all the leaves and leaf scars.

Advantages:

1. Gives extra income by way of sale of heads.
2. The crop matures twelve to fifteen days earlier than the head intact method.
3. Seed yield is slightly increased.

Disadvantages: In this method, flowering shoots are decumbent and require very heavy staking, otherwise they breakdown very easily while inter culturing or spraying.

Stump with Central Core Intact Method

In this method, when the crop is fully mature in the first season. The heads are examined for trueness to type. Plants with off-type. Plants with off-type heads are removed and rejected. Then the heads are chopped on all sides with downward perpendicular cuts in such a way that the central core is not damaged. This is an improvement over stump method in that the shoots arising from the main stem are not decumbent. During the last week of February and until 15th March when the heads start bursting, two vertical cross-cuts are given to the head, taking care that the central growing point is not injured. In the absence of such cuts, the heads burst out irregularly and sometimes the growing tip is broken. The operation is completed by going around the field twice or thrice during this period.

Advantages:

1. Shoots arising from the main stem are not decumbent, hence very heavy staking is not required.
2. Seed yield is increased.

Disadvantages:

1. The chopped heads cannot be marketed.

Head Intact Method: In this method, when the crop is fully mature in the first season the heads are examined for trueness to type. The plants with off-type heads are removed from the field. The head is kept intact and only a cross-cut is given to facilitate the emergence of a stalk.

Advantages:

1. The removal of heads (stump method) or chopping of heads on all sides (central core intact method) is not required. This saves time and labour.
2. Very heavy staking is not required.

Disadvantages: The seed yield is slightly low as compared to stump, or stump with central core intact method.

Cultural Practices

Time of Sowing and Transplanting: The sowing time of different varieties should be so adjusted as to complete head formation by the end of October or first week of November, by which time the mean temperature falls to 10^{0}C or below; at this temperature the heads stand best for over-wintering. Early varieties like 'Golden Acre' should be sown from 10th to 25th July and transplanted when the seedlings are three to four weeks old, during the second fortnight of August. This sowing time must be strictly adhered to, as the crop from the early sowings have matured heads during September and starts pre winter bursting and bolting. It is thus liable to be much affected by frost and snow during winter. Moreover, due to high mean temperatures of September (20^{0}C) the heads get infected with bacterial stock rot, which sometimes is very severe. The late crop, planted during September does not form heads and bolts directly during spring, and the seed grower is not able to ascertain purity of the crop. Medium-late varieties like burpee's Sure Head, and late varieties like Drum Head, which take about 2 1/3 to 3 months to produce mature heads, should be sown during the second and first fortnight of June respectively, and transplanting finished by the first week of August.

The mean temperatures 22.5°C, 20°C and 14°C of August, September and October respectively, afford optimum requirements for growth and head formation. The late transplanted crop starts head formation during spring and continues up to June and usually does not produce seed stalks (Singh., 1959).

Method of Sowing Nursery: The seeds are sown in raised nursery beds in a manner similar to that of cauliflower. Source of seed and seed rate: Obtain breeder's/foundation seed from source approved by a seed certification agency. Mid season and late varieties – 375 to 400 gm per hectare and early varieties 600-750 gm per hectare.

Preparation of land for transplanting: Prepare the land to a fine tilth by repeated ploughing and harrowing, followed by levelling.

Fertilization: Cabbage grows satisfactorily only when the supply of organic nitrogen is liberal. For best results apply 50 to 60 tones of farmyard manure per hectare, at the time of land preparation. Since the supply of farmyard manure is very limited in the hills, only 25 to 30 tones farmyard manure per hectare, is usually applied. Apply 200 to 300 kg superphosphate and 90 kg of muriate of potash by drilling. Top-dress two doses of 75 to 100 kg ammonium sulphate at intervals of two to three weeks after transplanting the seedling. Give another dose of 200 to 250 kg ammonium sulphate as surface application at the time of seed stalk emergence during March.

Transplanting: Three to four weeks old seedlings are transplanted. Transplanting should preferably be done in the evening and the field irrigated immediately afterwards.

Spacing:

Late varieties - 60 x 60 cm

Medium varieties - 60 x 45 cm

Early varieties - 45 x 45 cm

Irrigation: Cabbage requires a continuous supply of moisture. Irrigate the crop as frequently as required. Heavy irrigation should, however, be avoided when the heads have formed. A sudden heavy irrigation after a dry spell may cause bursting of heads

Hoeing and Weeding: At least three weedings and hoeings till the end of October are essential. One weeding and earthing up during November and December and the second during March, when seed stalks have emerged, control weeds and also help in proper drainage during winter and thereafter.

Roguing:

- The first rouging is done at the time of handling the mature heads.
- The Second roguing is done before the heads start bursting. The loose-leaved poorly heading plants, and those having a long stem and heavy frame, must be rogued out at this state. It is highly undesirable to keep such poor plants in the seed plots.

Harvesting and Threshing: Cabbage starts seed stalk elongation from 10-20th March when the mean temperature rises to 10-13 C. Flowering and pod formation starts during the first week of April at mean temperature of 13-18.5C. From 15th April to 15th May, the crop is in full flush of flowering and fruiting. The ripening of pods commences by 15th June to 20th June and the harvesting continues up to second week of July.

At mean temperatures below 20C during June and July, the maturity of crop is delayed at least by a fortnight and the harvesting may continue up to July end.

To avoid shattering of seeds, the whole crop is harvested in two or three lots with sickles. Generally, the early plants are harvested first and when the pod colour in about 60-70 per cent of the rest of the crop changes to yellowish-brown it is harvested completely and piled up for curing. After 4-5 days, it is turned upside down and allowed to cure for another 4-5 days, in the same way. It is then threshed with sticks and sifted with hand sifters. After thoroughly drying the seeds they are cleaned and stored.

Knol-Khol

Knol khol was first described in 18th century as another cabbage of western Europe. Knol-khol is characterized by formation of turnip like knobs, which is used for human consumption before it becomes tough and fibrous. This knobs arises from the thickening of the stem tissues above the cotyledons.

Time of Sowing and Transplanting: The seed is sown in the nursery from seventh to fifteenth August the seedlings are planted in the field during the first fortnight of September and the operation may continue up to the end of the third week the crops planted during October and later, fail to form good knobs.

Preparation of Land: For transplanting Prepare the field well by ploughing and three to four harrowings before the seedlings are set in the field.

Seed Rate: Main season and late varieties – 375 to 400 gm per hectare and early varieties 600-750 gm per hectare

Sowing of Seeds in Nursery: The seeds are sown in raised nursery beds in a manner similar to that of cauliflower.

Fertilization: Apply 20- 25 tones of farmyard manure per hectare at the time of preparation of field apply 100 to 150 kg superphosphate per hectare at the time of land preparation. Top-dress 100-150kg ammonium sulphate per hectare after first weeing. Top-dress another dose of 100-150 kg ammonium sulphate per hectare during spring when the crop recommences growth after over wintering. Extra application of nitrogen may be done before flowering, if necessary.

Transplanting: Three to four week old seedlings are transplanted. Transplanting should preferably be done in the evening and the field irrigated immediately afterwards.

Spacing:

Row to row - 60 cm

Plant to Plant - 45 cm

Intercultural: One hoeing and weeding during September to October, and one weeding and earthing-up during November to December is required. Keep the crop clean till the spring when one more hoeing and earthing -up is done (Verma and Sharma, 1999)

Roguing:

- Selection of Knobs is done during February to March when the knobs are well developed only true-to-type plants are retained. Off-types, or diseased plants, are removed.
- Subsequent roguing is done at the flowering stage. The off-type plants observed at the flowering stage, are usually determined by the extent of branching of the flowering shoots. the higher the production of fully branched plants, the greater is the seed yield. Therefore, care is necessary in nucleus seed production to select the right type of plants at flowering stage. Remove any undesirable plants.
- Subsequent roguing for off-types, diseased plants affected by diseases such as phyllody, black-leg, black rot, soft rot or leaf spot should be done from time to time as required.

Harvesting and Threshing: Seed stalk elongation starts from 10-20th March when the mean temperature rises to 10-13 C. Flowering and pod formation starts during the first week of April at mean temperature of 13-18.5C. From 15th April to 15th May, the crop is in full flush of flowering and fruiting. The ripening of pods commences by 15th June to 20th June and the harvesting continues up to second week of July. At mean temperatures below 20C during June and July, the maturity of crop is delayed at least by a fortnight and the harvesting may continue up to July end. To avoid shattering of seeds, the whole crop is harvested in two or three lots with sickles. Generally, the early plants are harvested first and when the pod colour in about 60-70 per cent of the rest of the crop changes to yellowish-brown it is harvested completely and piled up for curing. After 4-5 days, it is turned upside down and allowed to cure for another 4-5 days, in the same way. It is then threshed with sticks and sifted with hand sifters. After thoroughly drying the seeds they are cleaned and stored.

Seed Yield: The average seed yield is about 500 to 600 Kg per hectare.

Brussels Sprouts

Brussels sprouts is a popular vegetables in European countries which gained importance in the nineteenth century. The sprouts are the edible part of the plant, which are used, in different preparation. The sprouts are the buds in the axils of the leaves, which developed into miniature cabbage like heads. In India Brussels sprouts as not been common among the people. Like all other cole crops, it has been originated in the Mediterranean region from wild cabbage. It is a temperate vegetable requiring vernalization for transformation to the generative phase. Being biennial in nature, its seed production is possible in temperate regions, which are normally suitable for cabbage and knol-khol seed rising.

Floral Biology: Transition from vegetative to reproductive occurs only when it is exposed to low temperature of 2.22 – 100C for 42-56 days. The juvenile phase is complete in about 90 days. The flowers stalk arise from the terminal growing point and also from sum of the upper sprouts. The flowering takes place from the April to May. The flowers are typical of crucifers family having 4 petals 6 stamens and superior ovary. The anthesis of flowers depends on temperature and open mainly during hotter part of the day. Honeybees are the usual pollinating agents.

Cultivars

Dwarf Cultivars: Improved Long Island, Early Morn, Dwarf Improved, Frontier Zwerg and Kvik.

Tall Cultivars: Eveshan rapid, Wilhelmsburg, Hilds ideal, Red vein, Amager and De Rosny Polarstjernen

Climate: Brussels sprouts does well in cool and humid climate. This can be grown in regions with moderately severe winters. Although yields are higher in regions with cool mild summer and mild winters.

Plants at this stage of 30 leaves, the apex increases in size and becomes globular in shape due to the formation of sessile leaves and becomes globular in shape due to the accumulation of nutrients. At this stage the plant is sexually mature to receive the vernalization stimulus. On exposure to low temperature at this stage, the flower primordial will develop. As soon as temperature starts rising in spring, these plants bolt and produce flowers. Day length does not have any effect on this process. The sprouts become loose at higher temperature in autumn.

Soil: The soil should be moisture retentive and well supplied with organic matter. Waterlogged soil should not be used for seed crop, which results in poor and stunted growth of the crop.

Fertilization: Brussels sprouts need heavy intake of nutrients for the formation of firm sprouts of good quality, an even growth ism important. The recommended fertilize dosage is 200:80:100 kg NPK/ha. Nitrogen is applied in 4 equal split doses, first at the field preparation along with P and K, second 30 days after transplanting and the third at initiation of sprouts and the last at the time of bolting.

Spacing: Varieties with large leaves and larger sprouts are spaced widely at a spacing of 90 x 90 cm, whereas those with smaller leaves and smaller sprouts required a spacing of 60 x 60 cm or 60 x 45 cm. About 500-600 g of seed is sufficient to produce seedlings for 1 ha.

Sowing and Transplanting: The optimum time for sowing in the hills is second week of July and transplanting should be done by the end of August for seed crop. Strong winds during flowering and seed maturity stage do greast damage due to lodging because of the more height of the plant. The staking of plants is needed under such conditions.

Roguing:

1. Vegetative / pre marketable stage
2. Full grown stage
3. Bolting and pre-flowering stage

Seed yield: 300-400 kg/ha.

Sprouting Broccoli

Sprouting broccoli is less important in India and is grown by some people in the kitchen garden. It has been originated in Italy, as the name broccoli as been derived from Italian world 'brocco' refers to development of young shoots, which have been used as vegetables. Inspite of morphological resemblance with cauliflower and heading broccoli, the sprouting broccoli differs from these in plant habit and in producing green heads consisting of green buds and thick, fleshy flower stalks.

Floral Biology

The central growing point along with the axillary buds clusters form thick, fleshy flower stem, which is the actual consuming parts in broccoli. The length of the flower stem is small and the flower primordia develops into normal buds, which are united into a cluster, commonly called as head. Heads of broccoli becomes loose quickly and the buds start opening. The process of flowering starts at low temperature but pollination and seed setting will not occur until the average daily temperature goes above 150°C. Flowering starts from down upwards in all the branches, first at the main axis. The flowering is even and the whole crop flowers simultaneously. Between 20-300 C day temperature, flowering, pollination and seed setting go on satisfactorily. Higher temperatures result in pollen sterility and low seed setting.

Climate: It is not much sensitive to hot weather. A period of comparatively low temperature is not as essential for the production of sprouting broccoli seeds as it is for cabbage. Above mean temperature of 200°C, the flower buds open prematurely and the bud cluster become loose quickly, and results in undesirable leafiness in the heads. Sprouting broccoli is hardy and exhibits resistance to fairly heavy frosts depending on the variety and the developmental stages of the plant.

The annual varieties are more sensitive to frost in the vegetative stage than the biennial ones. But, in the bud production stage even a light spell of frost does a considerable damage resulting freezing, browning and ultimately rotting of bud clusters.

There are two types of seed production:

1. *Annuals:* are early whose seed can be produced in tropical / subtropical
2. *Biennials:* late in maturity although require same climatic conditions as are required by other Cole crops.

The period from sowing to flowering is shorter, so, they are sown slightly later than Cauliflower. Growing biennial type for seed is easier as they can withstand frost better in vegetative stage and over winter in the field. The annual types are not hardy and are susceptible to frost hence, the seed production is not possible in hills.

Method of seed production is exactly same as in Cauliflower since the transplanting method is not successful because there is no rest period between the sprouts/curd stage and flowering in broccoli and cauliflower. The transplanted allow to grow, over winter, flower and produce the seed at the same place i.e., in-situ method.

Manures and fertilizers:

10-15 tons FYM/ha

200:100:100 NPK kg/ha.

Nitrogen is applied in four splits, one is along with the P and K as basal dose, second after 25-30 days after transplanting, third 30 days after second application and last application before flowering.

Spacing:

Annual or Early varieties: 45 x 45 cm

Biennial or Late varieties: 60 x 45 cm

45 x 45 cm

Seed rate:

Annual or early varieties: 600-750 g/ha

Biennial or late varieties: 500-600 g/ha

Time of sowing should be adjusted in such a way that he heads are formed only after the danger of frost or snowfall is over. As the period from sowing to heads formation is shorter than cauliflower, sprouting broccoli should be sown or transplanted slightly later. Sowing of seed is done in third weak of October so that plants are well established to withstand frost or snowfall during winter. Being shallow rooted like other Cole crops, it needs shallow cultivation to check the weeds at the initial stage.

It can also produce seeds in the plains of Northern India, as seeding characters is similar to mid-season group of cauliflower. Since, it requires same temperature range for growth, head formation, flowering and seed setting and is not much sensitive to hot climate. Flowering period is shorter than Cabbage but equivalent to cauliflower. The adverse climatic conditions affect, flowering, pollination, fertilization and seed setting.

Rouging:

1. Marketable stage
2. Marketable size
3. Bolting and pre flowering stage

Harvesting: Generally the seed ripen in June, when 70% of the pods have turned brown and the rest of the crops change to yellow brown. Harvesting should be done preferably in morning hours to avoid shattering.

Seed yield: 500-600 kg/ha

Kales

Kales is of not much importance in India except grown by amateur kitchen gardeners. Like other Cole crops this has originated from wild Cabbage through mutation. There are different sub varieties, which bear a resette of leaves at the top of the stem.

1. Subvar. Laciniata: curly kales having finely curled foliage.
2. Subvar. Plana: smooth leafed kales
3. Subvar.millecapitata: thousand head kales with sprouted axillary buds forming loose leafy rosette

4. Subvar. Palmifolia; tree kales having more plant height up to 2 m.
5. Subvar. Medullosa; tree kales marrow stem kales with long swollen stem.

It is the curly kale that is grown for human consumption and is one of the earliest forms of Cole crops cultivated by the Greeks as early as 600BC. The top leafy rosette or individual leaves are harvested for vegetables. The other types of kales are used as fodder for livestock. For vegetable purpose kales should be grown at temperature around 18-200C as there is likelihood of the plants going to flower early. The flowers of kales are white and they retain their edibility even at the bolting stage.

Climate

For seed production kales behave as biennial in that it requires low temperature for vernalization. Kale is hardiest of Cole crops. Adult plants of hardy varieties can tolerate temperature from –100C to –150C. Freezing improves the quality of the leaves and even making edible the older leaves which have stringy midrib.

Seed Production

The seeds of kales can be produced in a manner similar to Brussels sprout. The crop continues making rosette of leaves at the top of stem till late winters. If the rise of temperature they start bolting and thus there is no dormant stage while transition in from vegetative to reproductive phase. Hence the replanted methods Is not successful and the seed is produced by insitu method, allowing the plants to grow, over winter, flower and produce seed at the same place.

Manures and Fertilizers

Nitrogen is applied in 4 equal split doses at different stages of crop at the final field preparation before transplanting, 25-30 days after transplanting, 50-60 days after transplanting and at the bolting.

Spacing:

1. For dwarf: 60x60 cm, 60x45 cm
2. For medium tall: 75x60, 75x45 cm
3. For tall: 90x60 cm

Seed rate

For dwarf: 500-750 g/ha

For medium tall: 400-500g/ha

For tall: 350-400g/ha.

Kales are transplanted late in the field during rainy season (Aug-Sept) like other Cole crops, so that the plants grow and over winter during the winter, flower in April and mature in June before the onset of rains. As the first season growth is confined to the production of only the leaves that is not particular regarding adherence to the proper time schedule as is the case with other Cole and root crops for the production of respective economic parts. Late planting results in reduced seed yield due to smaller plants size. Seedlings for transplanting are raised by sowing the seed 4-5 weeks ahead for transplanting.

Before the start of the winter, the plant should be earthen up to give them support. Bolting starts in March and flowering occurs in April. The bolting takes place in response to low temperature during the winter. The inflorescence is like other Cole crops excluding cauliflower and sprouting broccoli but resembles more closely to Brussels sprouts and knol-khol where in it develops directly from the rosette of leaves at the terminal end of the stem. Flowering and seed setting take place satisfactorily between 200 and 300 C. Pollen sterility and low seed setting result at high temperature. Harvesting: the crop is harvested when 70% pods have changed to yellow but before they are dried. Seed yield: 300-400kg/ha.

Hybrid Seed Production in Brassica Oleraceae

Because of Self-incompatibility it is a barrier to effective self-fertilization, under field conditions, out crossing is the rule and any given open pollinated variety is highly heterozygous. Inbreeding through bud pollination, usually results in complete loss of vigour in competition for survival.

Self-Incompatibility: The mechanism providing individual isolation is self-incompatibility, which means that pollen of a given plant will not fertilize the ovules of that plant, but is fully effective in fertilizing the ovules of most other plants of the same variety. Sporophytic incompatibility is the operating system in Brassica for the production of F1 seeds. This method involves two cross compatible lines, each of them are highly self-incompatible. The lines are individually propagated by bud pollination or open pollination at elevated temperatures first by selfing and then by sib pollination in isolation. In hybrid seed production block these lines are planted together in alternate rows to cross-pollination each other. The hybrid seed is harvested from both the parents (Bose et al., 2000).

Seed Processing

Harvesting and Seed Extraction

Harvesting is one of the agronomic management practices that require technical knowledge on maturation of the crops. This knowledge is much important in seed production than in commercial production.

1. Advantages on correct method of harvesting
   - Seed quality will be more
   - Seed yield will be protected without loss due to shattering
   - Processing loss will be reduced
   - Seed storability will be more
   - Above aspects will improve the seed marketability
2. Physiological maturation (PM)
   - The correct stage of harvesting for seed crop is termed as physiological maturation i.e., only after the attainment of physiological maturation crop should be harvested.
   - Physiological maturation is the stage at which the dry matter accumulation of the seed will be the maximum and it will be associated with maximisation of seed germination and seed vigour. The moisture content of the seed will be in decreasing order.

Physiological Maturity Characters: Physiological maturation is the stage at which seeds have maximum dry matter, germination and vigour.

Moisture content:

- The attainment of physiological maturation is represented by sigmoid curve of growth pattern.
- The physiological maturation can be represented both as duration and visible symptoms.
- Physiological maturation during is the days taken for maturation after the anthesis / flower opening / fertilization.

Fixing the days and harvesting is difficult. Hence for physiological maturation is identified based on visual symptoms that occur in fruit and seed characters.

The Visual Symptoms are: Physiological maturity symptoms in different crops

Crop Symptoms of Maturity:

Brinjal: Yellow colouration of fruits

Tomato: Reddish colouration of fruits

Bhendi: Hairline crack formation in dried fruits

Ashgourd: Ashy coating over the fruit

Pumpkin: Yellowish browning of fruit rind

Ribbed Gourd: Rattling sound of seed in fruit. Browning of dried fruits.

Snakegourd & Bittergourd Change of fruit colour yellow / red at distal end or any part of fruit. Onion Umbel colour change into yellow colour (10 – 20% - Black seeds visible) Caution in this visible symptom is sometime due to insect attack the visible symptom will be obtained with fruits earlier to physiological maturation

- These symptoms are easy to identify even by illiterates and alarms for harvesting crop with good seed quality. Harvesting the crop of physiological maturation is difficult at physiological maturation based on days is calculated more or less based individual seed and physiological maturation based on visible characters, all fruits of a population will not come to physiological maturation storage at single phase which will be more common in plants with indeterminate flowering habit than with determinate type. The solution for the problem is harvesting of crop at harvestable maturity. George (1980) divided vegetable seed crops into three broad groups, depending on the state of seed at harvest time.
    1) *Dry Seed:* seed is usually dried on the plant before harvesting eg: brassica, peas, beans onion
    2) *Fleshy fruits:* ripened fruits are picked form the plants and dried first, the dried fruits are then opened later to remove the dried seeds eg: chilli, okra, gourds pumpkin and eggplant
    3) *Wet fleshy fruits:* Fruits are harvested at high moisture level, the seeds has gelatinous or mucilaginous coating adhering to it. This has to removed after seed extraction by a fermentation process or treatment with dilute acids. Eg; Tomato and cucumber Methods of harvesting.
- Harvesting of crop can be done either mechanically or manually.
- Mechanical harvesting can be done only as single or once over harvest.
- Care should be at mechanical harvesting to avoid mechanical injury by adjusting blade size, gap between blades, speed of operation etc.
- Manual harvesting is done in two methods which is single harvest or periodical harvest.
- Either single or periodical harvest termed as picking is depending upon the growth habit of the crop. In determinate type where flowering is confined to a shorter duration once over or single harvest is recommended.
- In crops with indeterminate growth habit, periodical harvest or picking is recommended.
- Picking is harvesting the crop as and when a part of the population attains HM.

- The number of picking varies with crop depending on the growth habit of plant.
- Usually seeds of later picking are not considered for seed purpose.
- The type of harvest recommended for different crops is as follows.

Harvesting methods of vegetable corps

Amaranthus: A lightening or yellowing of foliage colour is an indication that seeds are beginning to mature. Types with an apical inflorescence are usually harvested once. Types with several side shoots are harvested several times as the seeds mature. The harvested seed stalks are placed on a clean tarpaulin or into very fine mesh nylon bags and allowed to dry in the shade.

Processing: Seeds are easily threshed by hand. The threshed seed is cleaned by winnowing.

Bean: Pods are harvested when they have turned yellow but are not yet completely dry. The inner seeds will be firm, well developed, and beginning to loosen inside the pods. Harvesting is often done in the morning to avoid losses due to shattering. Dry the pods in the sun and then place in a shelter for 1–2 weeks of curing. Pods are threshed by hand, being careful not to injure or split the seeds (injured seeds will germinate without primary leaves and grow poorly). Seed is further cleaned and dried after threshing.

Beet Family (beet, spinach and Swiss chard): Seed does not ripen uniformly on the plant and sheds easily when mature. Cut stalks when most flowering clusters have turned brown and stalks have turned yellow and dried—the first seeds are shed at this time.: Store stalks in a cool dry location for 2–3 weeks to encourage further seed ripening. Do not heap stalks on top of one another since this causes seeds to ferment. Handle stalks as little as possible since seeds shed easily. Small quantities of seed can be stripped by hand as seed matures. Large numbers of stalks can be put into a bag and beat with a stick. Chaff is winnowed away.

Brassicas (broccoli, Brussels sprouts, cabbage, Chinese cabbage, kohlrabi, mustard and turnip) Brassica seed shatters easily. Harvesting is done carefully when 60–70% of the pods have turned brown and most of their inner seeds are light brown and firm. Harvested seed stalks are cured for 1–2 weeks. Pods are then threshed with sticks and sifted by hand. Seed is brittle and should not be crushed when handled. Seed is dried in partial sun, then cleaned and stored.

Carrot: The seed turns brown 6 weeks after pollination. Before the seed shatters, cut and place umbels into paper bags to dry completely. Late-season rains will reduce seed quality. For small amounts, handpick each umbel as it dries brown. Large amounts of seed can be harvested by cutting the entire stalk as umbels begin to dry. Allow seed to mature in a cool, dry location for an additional 2–3 weeks. Seeds can be removed by hand-beating or rubbing umbels between hands. Winnow to clean. Remove spines from dry seed by rubbing.

Cucumber Family: The fruits should be left to fully ripen and turn colour. The fruits of luffa and bottle gourd should be left on the plant until they dry. For cucumbers, fruits will turn brownish colour. Bitter gourd fruits will turn orange. Some wax gourds will be covered with a pale-white powdery wax on the surface of the fruit. After harvest, the fruits can be kept in a shed for a couple of weeks to allow the seed to further ripen. For 'wet seeds' such as cucumber, wax gourd, bitter gourd and melons, cut the fruit lengthwise and scrape seeds out with a spoon. Allow seeds and the jellylike surrounding liquid to sit in a container at room temperature for 1–2 days. Fungus may start to form on top. Stir daily. The jelly will dissolve and good seeds will sink to bottom while remaining debris and immature seeds can be rinsed away. Spread seeds on a paper towel or screen until dry. For 'dry seeds' such as luffa and bottle gourd, keep the seeds in the fruit until they naturally separate from the flesh. This can be identified when you shake the fruit, the sound of seeds moving inside is heard. Cut off the bottom of the fruit and shake the seeds out, winnow to clean the remaining chaff, then place the seed on a screen for further drying before storage.

Eggplant: Harvesting is done when fruits are fully ripe (the skin of fruit turns brownish-yellow in green varieties or brownish in purple varieties). Harvest and store the fruits in a shed for a week until the fruits get soft. The outer covering is peeled off and the flesh with the seeds is cut into thin slices. These are then softened by soaking until the seeds are separated from the pulp. If the material is allowed to stand overnight in this condition, the separation of seeds from the pulp becomes easier. After separation, the seeds are dipped into water. The plump seeds will sink to the bottom. The seeds should then be dried on a mesh for a couple of weeks in a cool, dry place before storage.

Processing to improve seed quality after the harvest, the fruits not confirming to the varietal character, small fruits and pest and disease infested fruits must be removed and only graded fruits are to be used for seed extraction.

Seed extraction methods: The seed viability depends on the method by which the seeds are extracted and hence, it is more important to choose proper method of seed extraction. The selected fruits are to be cut into pieces and put in the cement tank with water for 10-12 hours for fermentation. The fruits can be allowed for over ripening for one or two days. This facilitates for easy pulping of fruits. Then fruits are to be made into pulp by manually. Add excess quantity of water and after ½ hour, remove the floating pulp fraction and collect the seeds settled at the bottom. For large scale seed extraction we can use the brinjal seed extractor.. The seeds extracted by this machine may again be treated with concentrated HCl @ 2-3 ml/kg seed with equal volume of water for 3-5 minutes with constant stirring and then seed should be washed with water for three to four times. It is easy to dry the seeds extracted by acid method and also remove the fungus growth over the seed coat, thus seeds possess golden yellow colour and high vigour.

Okra: The okra pods mature in a sequence from the base of the plant toward the top. The pods have tendency to split along the suture when they are dried out. Exposed seeds may be damaged by rain or may drop to the ground; therefore, the pods must be harvested as soon as they have become fully mature (brown colour) and before shattering. Pods are easily hand threshed.

Onion: Clip umbels as soon as most flowers have dried and before seeds begin shattering. Some growers harvest when the seeds are exposed in 10% of the umbels. Fully dried flowers will drop clean seeds naturally. For small amounts, rub remaining flowers to free seeds. For larger amounts, rub heads over screens. Winnow to remove remaining debris. Allow to dry in cool, dry location for up to 2–3 weeks. Artificial drying is often used.

Chilli: Harvest mature, fully-ripe peppers for seed. Most peppers turn red when fully mature. Pepper seeds may be extracted from fresh fruits or from fruits that have been dried in the sun for a few days. Seeds may be removed by hand or extracted by grinding the fruits and separating the seeds from fruits with a series of water rinses. Spread the seeds on a screen for drying under shade for 2–3 days but bring them inside every evening.

Radish: Threshing seeds from pods are difficult. Allow the plants to mature fully; then bring them to the threshing floor for drying. Threshing is done by hand-beating with sticks. The seed is then dried further.

Tomato: Allow tomatoes to completely ripen on the plant before harvesting for seed. Seeds from green, unripe fruits will be most viable if extracted after allowing the fruits to turn colour, but this is not advisable. Cut each tomato into half at its equator, opening the cavities that contain the seeds. Gently squeeze out the jelly-like substance that contains the seeds. Place the jelly and seeds into a small container for fermentation; add a little water if you are processing only one or two small tomatoes. Loosely cover the container and place in a warm place (around 25–30 °C) for 1–2 days, stirring daily.

A layer of fungus will begin to appear on the top of the mixture after a couple of days. This fungus not only eats the gelatinous coat that surrounds each seed and prevents germination, it also produces antibiotics that help to control seed-borne diseases such as bacterial spot, canker, and speck. After fermentation, fill the seed container with water. Let the contents settle and begin pouring out the water along with pieces of tomato pulp and immature seeds floating on top. Viable seeds are heavier and will settle to the bottom of the container. Repeat this process until water being poured out is almost clear and clean seeds line the bottom of the container. Pour these clean seeds into a fine-mesh strainer. Let the excess water drip out and invert the strainer onto paper towel, fine mesh, or newspaper. Allow the seeds to dry completely in an oven or in partial shade. Break up the clumps into individual seeds, label and store for later use.

1. *Fermentation method:* The ripe fruits are crushed well in a nonmetallic container to make slurry. Do not drain out the juice. Keep the entire material till ferment. When fermentation process completed profuse foam forms on the upper surface of the material and tomato flesh separated from seeds completely. The liquid is decanted off and seeds are washed at least for 8-10 times with clean water. Spread the seed on cement floor thinly and sundry.
2. *Alkali method:* Add 300 grams of washing soda to 4 liters of boiling water. Take 4 liters of extracted pulp and add the washing soda solution. Allow the mixture to stand overnight in an earthen pot. Next day all the seed will settle down at the bottom of the container. Decant the liquid. Wash the seed thoroughly with clean water and allow it to dry under sun.
3. *Extraction with Hydrochloric acid:* Add commercial Hydrochloric acid @ 5ml per kg pulp and mix thoroughly by stirring and leave for 30 minutes. Again stir the mixture and wash the mixture with clean water repeatedly for 8-10 times and allow the seeds to dry under sun.

Dry the seeds up to 8% moisture. Treat the seed with Thiram or Captan 75% WDP @ 2grams per kg seed and store the seeds in a moisture proof container which will help in retaining 80% germination at the end of 2 years.

Cowpea: Allow pods to dry brown before harvesting. The first harvest will be 8–10 weeks after sowing, followed by 2–3 harvests per week during the 6–8 week season. Cut pods with a sharp knife to minimise plant damage. Pods that are harvested 20 days after pollination will give the maximum seed quality. Pods are dried in the sun for approximately 3 days. For small amounts, pods may be opened by hand. For large amounts, hang the pods in a burlap bag and beat them with a stick, or put on the floor and walk on them. Remove large chaff by hand or winnowing. Discard blemished and shrivelled seeds. Place remaining seeds under shade for 1–2 weeks for further drying.

Seed Processing

Seed processing is to narrow down the level of heterogeneity of the lot by using suitable processing methods.

This heterogeneity can be narrowed down in the processing of seeds by eliminating the undersized, shrivelled, immature, ill filled seeds using appropriate sieve size. So if grading is done to obtain a particular range of size, shape, length and density of the seeds the quality of the lot is upgraded.

Requirements in seed processing

1. There should be complete separation
2. Minimum seed loss
3. Upgrading should be possible for any particular quality shape
4. Efficiency
5. It should have only minimum requirement

Physical characteristics used to separate seeds are

1. *Size :* Based on size it can be separated with air screen cleaner cum grader
2. *Length :* Disc or indented cylinder separator
3. *Weight :* Specific gravity separator
4. *Shape :* Spiral separator or draper separator for round and flat seeds
5. *Surface texture :* Rough from smooth surface seed- dodder mill
6. *Colour :* Electronic colour separator
7. *Electrical conductivity:* Seed differing in their ability to conduct electrical charge can be separated with electronic separator.
8. *Affinity to liquid :* The seed coat of seed will absorb water, oils etc., which provides a means of separating seed on the magnetic separator.

Seed Processing Equipments

I. Air Screen Cleaner: The dried seeds should be cleaned and graded with help of a Air screen cleaner (cleaner cum grader).

It consists of the following parts:

1. A hopper in the top for seed filling
2. A flutted roller below the hopper to regulate the seed flow to the screen.
3. Screen (or) sieves - Perforated metal sheet with specific size of perforation in which there are two types.
   i) Rectangular perforations for paddy and
   ii) Round perforations for seed other than paddy
4. Screen shaking unit: for oscillating the sieves to move the seeds on the screens
5. Screen brushes to remove the blocked seeds

6. Air blower with adjustments for air outlet
7. Collecting outlet
8. Air duct for directing the blown up light particles to outside
9. Collecting bins.

Working Principle: When seeds are fed into the hopper and they are guided to fall on the first sieve. The first sieve is a scalping screen which scalps all the foreign materials larger and heavier than seed and the entire quantity of seed passed through the first sieve.

The second sieve is a cleaning sieve which removes all the unwanted particles larger in size than the seed. The third sieve is actually the grading sieve which size grade the seed lot and bring into a uniform size and which also screens the undersized, shrivelled, immature seed, dust and dirt. The seeds are then rolled and passed through air column, where they are relived of the light chaffy and other materials by the blowing air.

Adjustments

Flutted Roller: The speed of this roller can be adjusted so as to increase (or) decrease the flow of seeds the hopper to the sieves.

Slope (or) inclination of the screen

The angle of inclination of the screens can be adjusted according to nature of seeds.

Rate of vibration of sieve

This can be adjusted either to increase or to decrease the speed of the rolling seeds on the screen.

Volume of air flow

By increasing (or) decreasing the air inlet.

Choice of screens

According to variety we have to change the screen

Screen dams

Small check dams, which can be provided here and there on the screens so that the seeds can be stopped a while and takes the charge either to pass or to roll.

Types of seed cleaner cum grader

i) Crippen model cleaner cum grader
ii) Clipper model cleaner cum grader
iii) Petkas cleaner cum grader

II. Disc Separator: It consists of a series of discs, which revolve together on a horizontal shaft inside the cylindrical body. Each disc contains many under cut pockets. The seed enter the intake end of the separator and move through the open centres of the discs towards the discharge end of machine. As the discs revolve through the seed mass the pockets lift out short seed but rejects longer seed. Longer seeds are conveyed by flights on the disc spokes towards the discharge end of the machine where they go out through the tailings gate. The rate of seed travel through the open disc centres is controlled by conveyor or blades attached to the spokes of the discs. The discs separator makes a very precise separation. No factor other than seed length and shape affects its separation. Flexibility is obtained by varying the size of the pockets.

III. Indented Cylinder Separator: The indented cylinder separator is a rotating, almost horizontal cylinder with a movable, horizontal separating adjustments which are mounted inside it. Indents line are there inside the surface of the cylinder. The indented cylinder revolves ,turning the seed mass to give each seed a chance to fit into indent. Short seeds are lifted out of the seed mass and are dropped into the lifting through long seeds remain in the cylinder and are discharged out via., a separate spout at the end of the cylinder.

As the cylinder revolves, it creates centrifugal force which helps to hold the seed in the indent. Short seeds are held in the indent until the cylinder turns to the point where the indent is inverted enough for gravity to cause the seed to fall out of the indent. The length, surface texture and size of seeds all determine how they fit into the indent, so that it can be lifted out of the seed mass. The speed of the cylinder creates centrifugal force which holds the seeds in the indent as it is lifted upward. Thus the shape and size of the seed to cause some seeds to fall out after being lifted only a short distance, while other seeds are lifted closer to the top of the cylinder before they fall out.

As the seeds enter the cylinder, the small, short, easy to separate seeds are quickly removed. The centre cylinder section removes the intermediate sizes of seeds still in the cylinder. All indents in a cylinder are the same size, only the progressively declining amount of material to be lifted causes this difference in separating action.

Adjustments

1. Cylinder speed
2. Size of the indent
3. Through setting
4. Tilt of the cylinder
5. Adjustable retarder.

IV. Specific Gravity Separator: Seeds of the same size and general shape can often be separated because they differ in specific gravity or relative weight. This difference is very useful in removing light, immature seeds or heavy sand and rocks to improve the purity and germination of crop seeds. If seeds which differ in specific gravity (relative weight / unit of volume) are placed on substrate of intermediate density, seeds of higher specific gravity will fall down through the substrata, while seeds of lower specific gravity will be buoyed up the substrata. Here air is used as a separation substrata.

As seeds flow on the deck of the gravity separator, they enter a column of air coming up through the porous surface of the deck. The pressure of terminal velocity of the air rising through the deck can be controlled very closely to separate two kinds of seeds differing in specific gravity, the air is adjusted so that only the lighter seeds are lifted up off the deck surface. These lighter seeds are held up by air pressure and tend to float on the deck surface. The heavier seed possess a velocity greater than that of the air column's so they are not lifted and so will lie on the deck surface. The air column thus stratifies the seed mixture into vertical zones of relative weight with the heavier seed lying on the deck and the lighter seeds lifted up to the top of the seed mass.

Adjustments:

1. Feed rate
2. Air flow
3. End slope
4. Side slope
5. Deck oscillation speed
6. Deck speed.

Seed Treatments

Based on the purpose of seed treatments it can be classified in to two groups

A. Pre - sowing seed treatments

B. Pre-storage seed treatments

A. Pre Sowing Seed Treatments: It is the treatments given to the seeds before sowing to improve the germination and vigour potential and as well as to improve the health of the seed. Pre sowing seed treatments includes the following:

a) Dormancy breaking treatments

b) Chemical treatments improve germination and vigour potential.

c) Insecticidal and fungicidal treatment.

d) Accelerating the speed of germination treatments (other special treatments)

   a) Dormancy breaking treatments: Refer seed Dormancy chapter

   b) Chemical treatments to improve germination and vigour potential. Soaking / treating the seeds with nutrients vitamins and micronutrients etc.,

   c) Insecticidal and fungicidal treatment.

Seed health : It is an important attribute of quality seed. A seed lot that meets high standards of germination, vigour, and purity but if it is contaminated with seed borne pathogens and insect pests, may be useless to farmers because it may result in severe yield loss or even crop loss in an entire area.

Benefits of the Insecticidal and fungicidal treatments.

1. Prevents the spread of plant diseases
2. It protects the seed from seed rot and seedling blights.
3. It improves the seed germination
4. It provides protection from storage insects.
5. It controls the soil insects.

Types of fungicidal and insecticidal seed treatments, they are 3 types they are

1. Seed disinfection
2. Seed disinfestation
3. Seed protectants

   1. *Seed disinfection:* It refers to the eradication of fungal spores that have become established with in the seed coat (or) in more deep seated tissues. For effective control, the fungicidal treatments must be penetrate the seed in order to kill the fungus that is present.
   2. *Seed disinfestation:* It refers to the destruction of surface borne organisms that have contaminated the seed surface but not infected the seed surface. Application of chemicals through chemical dips, soaks, fungicides applied as dust, shiny or light have been found successful.
   3. *Seed protection:* The main purpose is to protect the seed and young seedlings from organisms in the soil, which might other wise cause decay of the seed before germination.

Formulation of Fungicidal Materials/Insecticidal: Fungicidal / insecticidal seed treatment materials are available in the form of dusts, wettable powders and liquids.

1. ***Dusts :*** It is usually applied @ 200-250 gms / quintal of seed. Main disadvantage is dusty condition will prevail during the seed treatment and after handling.
2. ***Slurry :*** This type of fungicide is applied to the seed along with soap like water suspension which can be mixed with seed by using special slurry treater.
3. ***Liquids :*** The use of liquid solution is known as the "quick wet ' method. Here a volatile fungicide is applied to the seed and it throughly mixed with them.

Accelerating the speed of germination treatments (other special treatments)

i) Seed hardening treatment: Seeds can be hardened for 2 purposes I) Drought tolerance ii)

Cold tolerance: Seeds has to be imbibed in solution. Once the seeds imbibes water, the germination process takes place. At the end of soaking period the seeds should be dried back to its original moisture content. Chemicals used : $CaCl_2$, KCl,

ii) Seed fortification: Main aim is supply nutrients to seeds. The main objective is to have the high vigour, to over come unfavourable soil reactions. eg.) seed fortification with $MnSO_4$ @ 0.5 to 1 %. will improve oxidation - reduction potential of seeds, which ultimately leads to higher germination.

Seed pelleting: Here the nutrients are coated on the seeds. This technique is very much adopted in forest tree seeds.

Importance

- Normally in small seeds this technique is adopted.
- By pelleting we can increase the size of sees and we can make it free flowing one.
- Through this we can able to reduce the seed rate.
- It is also important for aerial sowing (gum arabica) in tree seeds.

Materials Used : Nutrients , adhesive, filler material.

Inert Materials: Lime, $CaCO_3$, Chalk powder.

Seed Sampling

Seed Sampling: It is the process of obtaining a sample of required size for test in which the same constituents are present as in the seed lot and in same proportion. Types of samples received at STL, Service sample, Official sample, Charged sample, Seed sampling. Primary Sample: a fraction of seeds taken randomly from seed lot different places.

Composit Sample: pooled sample of all primary samples.

Submitted Sample: sample which is submitted by producer to seed testing laboratory. It is large in size.

Working Sample: subdivision of the submitted sample.

Equipments for Sampling: SeedTriers-Stick or sleeve type trier, Bin sampler, Nobbe trier

Seed Dividers: Boerner seed divider, Gamet seed divider, Soil type seed devider

Random Cup Method: Suitable for crops requiring sample up to 10g. Place 6-8 cups in the tray and pour seeds uniformly over tray and select the cup randomly.

Spoon Method: Permissible for sampling of small seeded seeds. Pour the seeds evenly over the tray. Remove small portion of seeds using spoon and spatula.

Hand Halving Method: Restricted to specific genera of chaffy seeds. Pour the seeds on floor and make a mound. Mound is then divided into half and each half is halved again.

The primary objective of purity analysis is to determine whether the submitted seed sample confirms to the prescribed quality standards in regard to purity components, objectionable weed seeds, other inseparable crop seeds and seeds of other distinguishable varieties. Equipments and Materials: Purity table or working board, Seed divider, Forceps, Spatula, Brush. Aluminium purity dish, Magnifier (5 to 7x), Analytical balance, Hand screen (with ISI specifications), Top loading balance (as mentioned above), Stereoscopic binocular microscope, Seed blower, Table lamp, Watch glass

Components of Seed

Pure seed: Seeds of kind/species stated by sander or found to predominant in the test. Other crop seeds: seeds of plants which are as crops , other than main crop. Inert matter: includes seed units and all other matter and structures not defined as pure seed, other crop seed or weed seed. Weed seeds: Seeds of a weed species which are recognised as weeds by law/general usage.

Reporting of results:

- Weight by percentage Single decimal place
- All components should add to 100 %
- Less than 0.05% reported as trace
- Any component is found to be nil- reported as 0.0
- The components scientific names should be mentioned

Germination

It is the emergence and development from the seed embryo of those essential structures which for the kind of seeds being tested, indicate the ability to develop into a normal plant under favourable conditions in the soil

Objectives of the germination test

- To know the field planting value of the seed
- Results can be used to compare among the seed lots

Materials Required

Sand (0.05 to 0.8 mm particle size)/Germination paper, Germinaton chamber, Plastic boxes for seed germination in sand, Rubber band Glass Marking Pencil, Miscellaneous laboratory glass wares

Different Substrata

Paper media:

- Germination paper
- Blotter paper
- Pleated paper

Sand Media

Characteristics of germination paper

- It should be porous in nature
- It should have maximum water holding capacity to ensure continuous supply of water during the test period
- Free from bacteria, dirt, fungi and toxic substances
- Made out of 100% cellulose with
- pH should be 6-7.5
- Characteristics of germination paper
- Paper should posses sufficient strength to the prevent penetration of root in to the paper
- Paper size is 46 X 29cm
- It should have reasonable cost
- Should not serve as suitable media for saprophytic Fungi

Methods of germination test

1. Between paper method
2. Top of the paper method
3. Sand method

According to ISTA 400 seeds, 100 seeds in each replication (4 repns) has to be used for seed germination test. Spacing between seeds should be uniform and there should not be any over lapping of seeds Temperature to that specific species should be maintained. Species which required light (Grasses) has to be provided

Evaluation of germination test:

1. First count
2. Final count

Categories of Seedlings: Normal seedlings: It shows the potential for continued development in to satisfactory plant when grown in good quality soil and under favourable conditions of moisture, temperature and light.

Abnormal Seedlings: Don't shows the potential for continued development in to satisfactory plant when grown in good quality soil and under favourable conditions of moisture, temperature and light.

Fresh Ungerminated Seeds: Seeds which have not germinated by end of test period but remain clean and firm. Hard seeds: Remain hard by end of test period.

Dead Seeds: Seeds which are neither hard nor fresh nor produced any part of seedlings at end of test period.

Seed Moisture Test: The main purpose of moisture determination of seeds is to prevent loss of seed viability till it is planted for commercial purposes.

Seed Moisture Estimation

Rapid method using moisture metres hot air oven method rapid method using moisture metres:

- Electronic moisture metre
- Universal moisture metre
- Infrared moisture balance

Hot air oven method:

- Low constant temperature method (103 degree C for 17 hours)
- High constant temperature method (130 degree C for 2-3 hours)

Viability Test: Tz Test

Tz is a biochemical test and one of the quick methods to predict seed viability developed by Lakon (1942) in Germany.

Principle: In this biochemical test, living cells are made visible by reduction of an indicatordye. Indicator dye 2,3 5 Tripheny tetrazolium chloride (soluble, colourless and diffusible) interfeares with the reduction process of living cells and accepts hydrogen from the hydrogenases. Red, stable and nondiffusable Formazon is produced in living cells. This makes it possible to distinguish the red coloured living parta of seeds from the colour less dead ones.

Procedure: Preparation of seeds

1. *Bisect longitudinally:* eg. Sorghum, large seeded grasses
2. *Bisect laterlly:* eg.Small seeded grasses
3. *Remove seed cot:* eg. Dicots
4. *Conditioning only:* eg: large seeded legumes

Staining: Prepared seed should be placed in suitable container and cover with testing solution. Place the container in dark and worm place.

Evaluation: Should have good knowledge of staining pattern. If embryos and well stained thouse seeds are germinable.

Advantages of theTz test

- Rapid test
- More reliability
- Few equipments needed
- Disadvantages of theTz test
- More time required in seed preparation
- Skilled labours required
- Vague results

Seed Health Testing

The objective is to determine the health status of the seed sample i.e., whether the seeds are infected with fungi or not. Seed borne inoculum may give rise to spread of disease in field and reduce the commercial value of the crop. Further the seeds imported from out side source may introduce new pests and diseases into new areas.

Seed Health Testing Methods: Inspecting dry seeds Seedling symptom test, Agar plate method, blotter method etc..

Seed Vigour Tests

Seed Vigour; It's a sum total of a all the attributes which resulted in quick, uniform and early emergence of seedlings even in unfavourable environmental conditions The principle objective of a seed vigour test is to differentiate a range of quality levels, for example high, medium and low vigour seeds. Direct Vigour tests Indirect vigour tests Brick gravel tests (2-3mm size) Dry weight of seedlings Paper piercing test Speed of germination Seedling length measurement.

Application of Molecular Markers for Hybrid Purity Testing

Quality seeds has to meet the minimum seed certification standards and quality attributes viz., physical purity, germination per cent, moisture content, seed health and genetic purity. The genuineness of the variety is one of the most important characteristics of good quality seed. Genetic purity test is done to verify any deviation from genuineness of the variety during multiplication stages. For certification genetic purity test is compulsory for all foundation and certified hybrid seeds. Higher genetic purity is an essential requirement for the commercialization of any seed. Besides for the success of any hybrid technology depends on the quality of the seed supplied in time.

The genetic purity during multiplication stage plot is prone to contamination from presence of pollen shedders, out crossing with foreign pollen and physical admixtures due to careless handling during harvesting, threshing, processing and bagging. Using low genetically pure seeds will proportionately reduce the commercial yield. It is estimated that for every 1% impurity in the hybrid seed the yield reduction is 100 kg per hectare (Mao *et al.*, 1996). Maintenance of parental line purity is a prerequisite to ensure high genetic purity of hybrid seeds.

A certification standard of genetic purity for foundation is 99 per cent. While, for certified hybrids it is 95 per cent except cotton (90 %) and castor (85%). A set of qualitative and quantitative characters known as descriptors are currently in use for variety identification and description. Some of these characters, particularly those showing quantitative inheritance, interact with the environment in which the variety is grown and thus make the process of variety identification subjective.

Marker

Marker is a 'TAG' for identification. Identifying a genotype based on some recognisable mark can be used for its identification and characterization. It may be morphological or biochemical or molecular markers. Ideal marker should be polymorphic, Co-dominant, distributed evenly through out genome, heritable, independent to environmental factors and stage of the crop, easy, quick and cost effective to conduct and results must be reproducible

Methods to Assess Genetic Purity

Genetic purity of a given seed lot can be assessed by using one of the following methods

1. Conventional grow out test
2. Chemical test
3. Electrophoresis method

a. Biochemical markers (Proteins and Isozymes)

b. Molecular markers (DNA)

Molecular Markers

Molecular markers, in contrast, being based on DNA sequence variation, provide an unbiased means of identifying crop varieties. The Biochemical and Molecular Techniques group of the International Union for the Protection of New Varieties of Plants (UPOV) is evaluating different DNA marker parameters prior to its routine use in establishing distinctness, uniformity and stability (DUS) of plant varieties.

This is a new approach to test the genetic purity of the seeds done at nucleotide level. This test screen through whole genome and produces enough polymorphism in closely related parental lines also. DNA markers has advantages over morphological and biochemical markers. Those are the resultant on genotype and environmental effect. While, DNA markers are resultant of only genotype of the cultivar and are independent to environmental conditions. The markers are dominant and Co-dominant and can differentiate between homozygous and heterozygous conditions of the cultivar. This is rapid to conduct and results are reproducible.

The banding pattern appeared at a particular molecular weight is used for the identification of parental lines, hybrid and offtypes.

Importance - Stable Marker for Genetic Purity

CMS plants and its maintainer plants, which originated from female parent during multiplication, are major off-types in F1 hybrids. Maintenance of the purity of parent CMS lines is essential in achieving the purity of hybrid and thus commercial benefit. CMS purity during multiplication can only be assessed at heading stage by observing pollen fertility. The results are prone to be erratic due to different examiners and environments. So it is significant to develop a novel, simple, rapid and effective method to assess CMS seed purity during multiplication at seedling stage.

Hybridity Confirmation

Identifying breeding lines and determining hybrid purity are major requirements in plant breeding and seed production. To test the conformity of hybrid seed, one must be able to distinguish the true hybrid resulting from cross between the male and female parents and one coming from self pollination of the female parent. The DNA finger printing of parental lines, hybrid and offtypes are used as a data base to identify offtypes in the seed lot under question. Molecular markers which could clearly distinguish the hybrids form its parental lines have been identified and developed in major agriculture and horticulture crops are given in the following.

Researchers	***Year***	***Crop***	***Marker used***
Livneh *et al*	1990	Pepper	RFLP
Hu and Quiros	1991	Bracooli & Cauliflower	RAPD
Hashizume *et al.*	1993	Watermelon and tomato	RAPD
Ilbi	2003	Pepper	RAPD
Chin *et al.*	1996	Rice	RAPD
Rom *et al.*	1997	Plum	RAPD
Yang and Wang	2000	Rice	RAPD
Pendse et al.	2001	Cotton	RAPD
Chumg et al.	2002	Maize	SSR
Zhan	2002	Rice	SSR
Peng et al.	2003	Rice	SSR
Ashok and Vilas	2005	Cotton	ISSR
Joshi *et al*	2004	Bajra, Jowar	ISSR
Antonova et al.	2006	Sunflower	SSR
Namrata *et al.*	2007	Tomato	RAPD

The molecular markers are more efficient in assessing genetic purity. Among markers RAPD and AFLP are dominant markers. Dominant markers unable to identify heterozygous condition (AA and aa only but not Aa).

Low reproducibility of RAPD and lengthy process of AFLP markers have made them impractical and difficult for their routine use in seed purity analysis. While SSR, SCAR, STMS are C0- dominant markers and are able to identify heterozygous condition.

These markers are more popular because of their accuracy in results and are reproducible. Quick and simple process of these markers have made them practical for their routine use in hybrid conformity and seed purity analysis. Markers vary based on their ability to differentiate lines with the crop and hybrids and parental lines involved in developing particular hybrid.

Conclusion

Hybrid seed producer have to test every commercial hybrid seed lot produced. A fast and reliable method with capacity to screen large number of seed samples.

With the availability of high throughput PCR machines and simpler DNA extraction techniques may surpass the current methods of grow out test. DNA is now routinely used for the plant variety breeding purposes and it is likely that hybrid seed producers can adapt these modern techniques for hybrid seed purity analysis purposes.

Seed Marketing

Seed marketing is one of the vital component of seed technology. In broad sense it covers all the activities involved in the flow of seeds from production to consumption. However in narrow sense it refers to the actual acquisition and selling of packed seeds, intermediate storage, delivery and sales promotional activities.

Seed marketing comprised the followings

1. Demand forecasts
   a. Total cultivable area, seed rate, SRR, Seed multiplication ratio
   b. Impact of extension efforts
   c. Current acreage under high yielding varieties
   d. Farmers preferences
   e. No and size of compotators
   f. Kinds of publicity and sale promotional activities.
   g. Climate of the area where seed is being marketed.
2. Marketing Structure
3. Arrangement for storage of seeds
4. Sales promotional activities
   a. Print Media-News paper/news letters/Ads/ Posters/Banners
   b. Electronic Media-Cinima slides/Radio and TV/Mobiles/kisan call centres/Internet
   c. Krishi melas
   d. Field demonstrations
   e. Folk ways-skits/Mimes
   f. Project approaches-seed village concept
   g. Educational Institutions approaches-ATIC
5. Post sales service
   a. Technical help
   b. Quick follow-Up
   c. Attending the complaints
6. Economics of seed production and pricing

Seed Production cost:

a. All costs of crop production
b. Cost of foundation seed

c. Cost of specialised planting (Hybrids; male &Female sowing)
d. Cost of other special operations (Detaselling, emasculation)
e. Cost of additional supervision required in raising a seed crop
f. Seed certification fee
g. Other expenses: Construction of separate threshing yard or cleaning of floor and machineries.
h. Supervision of harvesting
i. Separate harvesting
j. Drying losses
k. Sorting of undesirable ears/heads/cobs
l. Additional transportation
m. Seed processing rejection
n. Seed processing and testing chargers
o. Seed marketing costs
p. Seed storage and transportation charges
q. Distribution and sales promotional costs

Seed Pricing:

1. Production, processing and storage cost
2. Distribution cost
3. Whole sale margin
4. Retail margin
5. Sale promotional costs
6. Risk costs
7. Over heads
8. Profit
9. Compotators price

Synthetic Seeds

Synthetic seeds are defined as artificially encapsulated somatic embryos, shoot buds, cell aggregates, or any other tissue that can be used for sowing as a seed and that possess the ability to convert into a plant under *in vitro* or *ex vitro* conditions and that retain this potential also after storage. In simple words synthetic seed contains an embryo produced by somatic embryogenesis enclosed within an artificial medium that supplies nutrients and is encased in an artificial seed covering.

The technology designed to combine the advantages of clonal propagation with those of seed propagation and storage. Also be as channel for new plant lines produced through biotechnology advances.

The first synthetic seeds were produced by Kitto and Janick in 1982 using carrot somatic embryos

Why Synthetic Seeds?

In some of the horticultural crops seeds propagation is not successful due to;

- Heterozygosity of seeds particularly in cross pollinated crops
- Minute seed size eg; orchids
- Presence of reduced endosperm
- Some seeds require mycorrhizal fungi association for germination eg: orchids
- No seeds are formed

These crop species can be propagated by vegetative means like micro propagation and clonal propagation.

Characteristics of synthetic seeds

1. High volume. Large scale propagation method
2. Maintains genetic uniformity of plants
3. Direct delivery of propagules to the field, thus eliminating transplants
4. Lower cost per plant let
5. Rapid multiplication of plants

Advantages of synthetic seeds over somatic embryos for propagation

1. Ease of handling while in storage
2. Easy to transport
3. Has potential for long term storage without losing viability
4. Maintains the clonal nature of the resulting plants
5. Serves as a channel for new plant lines produced through biotechnological advances to be delivered directly to the green house or field
6. Allows economical mass propagation of elite plant varieties

Based on technology established so far, two types of synthetic seeds are known:

1. *Desiccated:* The desiccated synthetic seeds are produced from somatic embryos either naked or encapsulated in polyoxyethylene glycol (Polyoxr) followed by their desiccation. Desiccation can be achieved either slowly

over a period of one or two weeks sequentially using chambers of decreasing relative humidity, or rapidly by unsealing the petri dishes and leaving them on the bench overnight to dry. Such types of synseeds are produced only in plant species whose somatic embryos are desiccation tolerant.

2. *Hydrated:* hydrated synthetic seeds are produced in those plant species where the somatic embryos are recalcitrant and sensitive to desiccation. Hydrated synthetic seeds are produced by encapsulating the somatic embryos in hydrogel capsules.

Procedure for Synthetic Seed Production: The somatic embryos foe synthetic seeds are produced in the lab through culturing of somatic cells and treating with different hormones to produce root and shoot. The following are the different steps involved in artificial seeds production;

1. Establish somatic embryogenesis
2. Mature somatic embryos
3. Synchronize and singulate somatic embryos
4. Mass production of embryos
5. Encapsulation of matured somatic embryos
6. Desiccation
7. Field planting

Somatic Embryos: Somatic embryos are bipolar structure with both apical and basal meristematic regions which are capable of forming shoot and root, respectively. Somatic embryogenesis is the development of embryos form vegetative cells with in vitro systems. Specific tissues have a capacity for somatic embryogenesis in cultural systems. This allows the clonal propagation of normally seed-propagated crops analogous to the production of apomictic seedlings. Somatic embryos develop through stages similar to zygotic embryos, however, the final size for the cotyledons are usually reduced and there is no development of endosperm or seed coat.

After pollination, a zygotic embryo of a dicotyledonous species develops through a series of morphological stages termed globular, heart and torpedo. Cotyledons develop and expand as the storage reserves of protein, starch and/ or oil are deposited. At some stages before the embryo achieves its maximum weight, it acquires the ability to tolerate drying. Then, the seed's vascular connections to the maternal plant are severed, it stops importing nutrients and it begins to lose water. Seeds of most crop plants can survive drying and can be stored for several years. Once they are hydrated, germination commences culminating in the emergence of a radicle and then the mobilization of the storage reserves by the seedling.

Zygotic embryos v/s Somatic embryos:

Zygotic embryos	*Somatic embryos*
Is the result of a sexual process	Produced as a result of asexual process
involve fusion of male and female gametes	doesn't involve male and female gametes
Produced form sexual cells	Produced from vegetative cells
contains genetic constituent form both parents	contains genetic constituent form single parent
Genetic recombination takes place	No genetic recombination will takes place
Contains embryo, endosperm and seed coat	Contains only embryo and endosperm and seed coat are absent

Procedure of somatic embryogeneis:

1) Petiole explants plants are surface sterilized and cultured on SH medium (Schenk and Hildebrandt, 1972) containing 2,4-D, kinetin and many other nutrients. 2,4-D activates the cell cycle of many cells in the petiole - those in the vascular cambium develop into a callus, whereas some sub-epidermal cells develop into a somatic embryo.
2) The initial somatic embryos, which are only small dense cell clusters at this stage, are embedded in a callus mass of non-differentiated cells.
3) To liberate these proembryonic structures, and to stimulate the formation of more embryos, the callus is dispersed in a liquid medium to form a suspension culture containing 2,4-D but not kinetin
4) After 7 days, the suspension is sieved and transferred to solid medium lacking 2,4-D On this medium the embryos develop through morphological stages that appear to be globular, heart and torpedo.
5) Maturation Phase 1: Once the majority of embryos reach the torpedo stage (7-10 days after sieving) they are transferred to an enriched medium containing a high level of sucrose, nitrogen and sulphur to prevent precocious germination and to enable deposition of storage reserves. The embryos rapidly accumulate fresh and dry weight, reaching 1-2 mg dry weight per embryo.
6) Maturation Phase Ii: To induce the acquisition of desiccation tolerance, the somatic embryos are placed on a modified medium containing abscisic acid (ABA) for 3 days. Then they are removed from the medium, washed to remove sugar and other nutrients, and dried.
7) The standard method of drying is to place the somatic embryos in a sealed chamber over a saturated salt solution designed to give specific relative humidifies. Daily for one week, the embryos are transferred to a progressively lower relative humidity chamber and finally are dried at ambient conditions. At this stage, the embryos have reached approximately 15% moisture and can be stored for a year or more with good viability.

Encapsulation of Matured Somatic Embryos

Somatic embryos produced naked embryos without storage materials and protective layer (seed coat).

This is very difficult for handling so this demand the encapsulation and coating. The somatic embryos produced are encapsulated using gel agents like agar, alginate, polyco, carboxy methyl cellulose, guar gum, sodium pectate etc.

Among these alginate encapsulation was found to be more suitable and practicable.

Alginate hydrogel is frequently selected as a matrix for synthetic seed because of its moderate viscosity and low spinnability of solution, low toxicity for somatic embryos and quick gellation, low cost and bio-compatibility characteristics. The use of agar as gel matrix was deliberately avoided as it is considered inferior to alginate with respect to long term storage.

Alginate was chosen because it enhances capsule formation and also the rigidity of alginate beads provides better protection to the encased somatic embryos against mechanical injury.

The somatic embryos are mixed with sodium alginate (2 %) and the suspension is dropped into the calcium salts solution (200mM).

The principle involved is when sodium alginate dropped into the calcium salt solutions it from round firm beads due to the ion exchange between Na^+ in sodium alginate and Ca^{2+} in calcium slat solutions and sodium alginate form calcium alginate in 20-30 minutes.

Since somatic embryos lack seed coat and endosperm the matrix of encapsulation can be added with nutrients and growth regulator, which will serves as an artificial endosperm.

This will increase the efficiency of germination and viability of seeds. Addition to these nutrients other useful materials fungicides, pesticides, antibiotics and micro organism can also be incorporated.

Application of Synthetic Seeds

By combining the benefits of a vegetative propagation system with the capability of long-term storage and with the clonal multiplication, synthetic seeds have many diverse applications in the field.

1) Multiplication of non-seed producing plants, ornamental hybrids or polyploids plants
2) Propagation of male or female sterile plants for hybrid seed production
3) Germplasm conservation of recalcitrant species
4) Multiplication of transgenic

Limitations

1) Limited production of viable micropropagules that are useful in synthetic seed production
2) Asynchrous development of somatic embryos
3) Improper maturation of somatic embryos that makes them inefficient for germination and conversion in to normal plants
4) Lack of dormancy and stress tolerance in somatic embryos that limit the storage of synthetic seeds
5) Somo clonal variations which may alter the genetic constituent of the embryos

5

Seed Production: An Overview

The production of marketable fish begins with the stocking of fry or juveniles into a rearing environment that assures optimum and rapid growth to allow harvest in the shortest possible time.

The fish farmer has to obtain adequate number of young fish to meet his production goals. These fish can come from wild capture.

However, there is little or no guarantee that adequate numbers can be captured and stocked in the time corresponding to optimum production conditions. The fish farmer then naturally turns to other means of obtaining his stock.

By simulation of the conditions necessary for the reproduction of his fish, the farmer can spawn the fish in captivity. Successful spawning is only the beginning, however, the eggs must hatch, and these reared successfully to fry stage. These stages - spawning, hatching, and early rearing are like a steeple chase which the farmer must win.

The race course is well filled with obstacles, for example physico-chemical quality of water such as available dissolved oxygen, feed of the proper nutritive composition and particle size, low resistance to diseases, and so on. A good appreciation of all these factors is needed for successful production of fish.

The ultimate goal of the fish farmer is to produce fish that meet both his needs and the market demand. Through artificial propagation, the farmer can select for desirable characteristics such as fast growth, resistance to disease, etc. By hybridization and selection, these goals can be achieved if the farmer dedicates enough time and patience.

Fish Seed in Aquaculture

As stated in the introduction, fish farming begins with the stocking of fry, and these can come from the wild or be produced on the farm. Whatever their origin, they are indispensable and the means of obtaining them influences directly farm production. If supplies are erratic, there will be interruptions in other farm activities; if the supplies are regular, farm production may be maximised. The cost of the fry can vary considerably and may be an important factor in overall production costs. It any event; a good supply of fry is essential for successful fish farming.

If one looks at production of eggs, larvae, and fry that is carried out on the farm itself, the major problems are obtaining a sufficient number of eggs, a good hatching rate of these eggs, and good survival and growth of the larvae obtained. In nature, there is very high mortality at these stages, and a lot of attention and effort is needed to overcome these difficulties.

To practice reproduction and fry production, a certain investment in equipment, infrastructure (ponds, tanks, water supply), and trained labour is needed. These costs can be considered a part of overall production costs of marketable fish. The proportion of total cost in producing saleable fish that is met by seed production should be kept always in mind, and efforts made to find new methods of seed production that are less expensive, hopefully increasing the profitability of the whole farm.

Definition of terms:

Larvae: hatchlings with yolk sack until first feeding.

Fry: free swimming fish - from first feeding until complete development of somatic organs.

Juveniles: sometimes referred to as "fingerlings", but generally fish of a small size but adequate for stocking, not sexually mature (Coche and Bianchi, 1979).

Methods of Seed Production

For certain farmed fish, it is not yet possible to understand and control all the stages of reproduction, and farmers depend on natural supply. In most cases, these fish are brackish water species and their biological cycle includes migrations into a different environment, either moving to the sea, or into fresh water. Thus to simulate the conditions needed for natural reproduction, this would mean a complete change in the environment. In general, fish can only be reproduced on the farm if conditions that correspond to natural spawning can be closely approximated. The cost of creating a completely marine habitat or freshwater habitat in the midst of a brackishwater situation would in most cases be prohibitive.

Collection of Seed from the Wild

Collecting eggs or fry from the wild was the first method used in obtaining stocking material. This is still occuring for species for which the spawning behaviour is not controlled or not well understood, or for which the costs of artificial propagation are too high or where fry in large quantities is easily obtainable. Means of collection differ with species. Three important species of fish will be examined.

Catfish: Throughout West Africa, the catfish (Chrysichthys spp) is very popular and in Ivory Coast represents an important species in brackishwater lagoon aquaculture. As it is relatively difficult to reproduce this fish in captivity, eggs are collected from the wild by placing bamboo traps in areas where this fish (C. walkeri) spawns. The traps are placed on rocky bottoms, in deep water (4 – 6 m), by fishermen and the catfish spawns inside the bamboo tube. This method allowed the collection of up to 2000,000 eggs even outside of the peak spawning periods. The eggs are transported to the hatchery, and placed in artisanal Zoug jars made from empty plastic bottles. Two or three days after hatching, the larvae are transferred into tanks made of plywood and covered with polyester resin measuring 2.4 × 0.4 × 0.3 m at a. density of 5,000 to 10,000 per tank.

After about 10 days, the yolk sac is resorbed. After two and half months, they reach an average weight of 750 to 800 mg. During this time, they are fed on a mixture of cattle brain and crab eggs, finely ground which forms a colloid suspension in the water easily accessible as food for all fish. After the first month, the feed is modified and composed of ground crabs, wheat flour (15% of the weight of crab) wheat bran (5%) and fish meal (20%). The fry are then transferred after 2½ months to circular concrete tanks with a running water system. The concrete tanks are 3 m in diametre with a central drain. After the 4th month, the fish are fed with a mixture richer in flour: crabs, fish meal 30%, wheat bran 20%, wheat flour 20%, copra cake 10%. Percentages are in relation to weight of crab. This feed is again changed during the 5th or 6th month to a feed composed of S! crab, and T! agricultural by-products (fish meal 35%, wheat bran 20%, wheat flour 25%, palm kernel cake 10%, copra cake 10%. The fish are then transferred to enclosures when they achieve an average weight of 4.5 grams. One fourth will have reached already a weight of 8 g, when half of the population will weight 4 grams, and ¼ 3 g. The fry are reared in enclosures until achieving a size of 50 g, which takes a further 6 months.

The obvious cost and time necessary for procurring catfish juveniles through this method is considerable, and further experimental work is necessary (Ledoux, 1979; Bailly, 1981).

Larger fish are sometimes captured from the wild. In Nigeria, the catfish are captured using hook and line with soap as bait. The catfish are injured, and care must be taken to avoid mortalities. It is easy to capture large numbers of catfish using traps to collect fish attracted to tilapia cages (Campbell, comm. pers.).

Milkfish: The milkfish (Chanos chanos) is found in warm waters of the Red Sea, and the Indian and Pacific oceans and is one of the most important brackish water species cultured particularly in S. E. Asia. In Africa, it is only found on the east coast. The fish reproduces once or twice a year in coastal marine waters of 25 m in depth. Each female spawns 1.5 to 1.7 million eggs, and sometimes up to 5 million. The larvae drift to the coastal waters near the estuaries with a temperature of about 23°C, the eggs taking about 24 hours to hatch. Salinity isn't too important (10 to 32 ppt) and the fish are captured where phytoplankton is abondant. When the fry have achieved 1.3 m in length, the yolk sac is completely absorbed and they begin looking for food. At this life stage, their diet is a mixture of blue-green algaes and the associated organisms (bacteria, protozoans). The diet will change later on. Artificial reproduction has been achieved in the laboratory, but not yet on a large scale, and thus almost all of the milkfish larvae are captured from the wild. For example, in 1970, in Taiwan only, 207 million fry were captured from the wild (Huisman, 1979).

According to the regions, capture techniques and material can differ. In Java, the best collection zones are on the north side of the island, and the fry are abundant 7 months out of the year. The best time for capture is at the spring tide, and the three days before and following the full moon. Sandy beaches, a slight slope, and clear, calm water are ideal.

The fish are often concentrated in the zone near estuaries where the influence of freshwater is high. They will often take refuge leeward of sand bars. If no sand bar exists, the fishermen will build a rock wall perpendicular to the beach into the ocean, or build artificial shelters using palm branches.

As the fry are collected, they are transferred into basins or similar containers, and the sea water is immediately diluted with freshwater. This will kill off many of the other organisms inadvertantly collected, and also acclimate the fry to brackishwater. Other species of fish are also captured, and these must be sorted out. This needs a certain level of expertise and experience on the part of the fisherman.

In the Philippines, the fry are captured by using seine nets or traps. In Mauritius, the fry are captured usually in July using a fine mesh seine net, and are transferred into acclimatization tanks. The mortality is usually high in the tanks (Bardach et al, 1972; Iversen, 1976).

Mullet: Mullet species are cultivated around the world, and have a large distribution. Mullets belong to the genera of Mugil or Liza, many species are known. The artificial reproduction is possible, however there are still many instances where fry available in important quantities is collected from the wild. The mullets reproduce in the sea, and will thrive well in the sea, in estuaries, in brackishwater, and sometimes migrate to freshwater. When they are sexually mature, they form schools, and the reproductive products are released into the sea where the fertilization occurs. The eggs are pelagic, and hatch in the following two days. After hatching, the larvae begin to swim near the coast where they usually arrive in mass after 2 months. At this point, they have reached a size of 25 mm. On the coast, they form small schools and begin to move towards the estuaries. The capture is usually done as they enter the estuary, at low or rising tide, using seine nets, deep nets, etc. The fish are euryhaline and have a good resistance to abrupt changes in salinity (Bardach et al, 1972; Iversen, 1976).

Shrimps: Among the shrimps used in aquaculture, there are two main groups, the Penaeidae and the Caridea. Their biology and life cycles are different. For an example of the Penaidae, Penaeus notialis is widespread and often farmed. Spawning is carried out in the sea and the female is recognised by well developed ovaries and colouration. She molts shortly before spawning. The male places a spermatophore on the female theca which is kept until egg deposition.

The various stages of post egg development occur in the sea. After roughly 3 weeks, the young, termed post-larvae, begin to migrate to estuaries. The larvae are attracted by light, and are often collected by seine nets and deep nets at night. In other instances, the farm is filled with water known to contain large numbers of postlarvae. Larvae can also be collected by using bunches of palm or coconut branches lashed together and placed on the bottom of the lagoon. After a few days, the traps are surrounded by a fine mesh net and the branches removed.

For the group of Caridea, the principal species raised is Macrobrachium rosenbergii. This species spends its life in brackish-water or freshwater, and migrates to the estuaries or lagoons for spawning. Either gravid females or post-larvae are captured. Again traps, deep nets, etc. are used for capture.

Problems with Natural Seed Collection Compared to Reproduction in Captivity: In certain cases, it is not possible to reproduce fish species in captivity however; but the capture of fry from the wild presents some problems. Sometimes the sorting out of species at the fry or fingerling stage is difficult, which will result in a mixed stocking population of different species - some fast growing, some slow growing, some predators and some competitors also

-into the pond. Identification keys for fry and fingerlings are not often available. Diseases and parasites affecting wild fish can be introduced into the pond. On the other hand, fry collected in nature have usually already passed the most critical stages of their life cycle where the mortality is highest, and have good survival rates when stocked. If reproduction and larval rearing is carried out on the farm, each problem at each step of the whole process must be solved. This may be expensive if infrastructure, equipment and skill is needed. For some species, as tilapia, however, not much is needed.

However, if using natural stocks, their abundance and demand will vary the price, and some years, the fry availability may be very low. Optimum collection periods will vary from year to year, and the ponds may not be properly prepared when the fry becomes available. In the two cases, it is important to have a sound understanding of the biology of the fish species either to duplicate reproductive conditions on the farm, or to know when and where to collect the fish. The necessary time spent in studying the fish, in collecting the necessary data on reproduction, etc, and then procuring the necessary equipment for capture may in the end be as expensive as artificial reproduction.

Incubation and Hatching

As soon as the egg is fertilized, embryonic development occurs inside the egg untill hatching. At this point, the larva tears through the shell and begins an independent life. During the egg development stages, the eggs are susceptible to predation, and attacks from bacteria and fungus growth. The time needed for hatching varies with the species, but in all cases is temperature dependent. In tropical waters where the water temperature is usually superior to 25 degrees, hatching usually occur in 1 to 5 days. In temperate climates, hatching can take several months in very cold water. Eggs can be categorized as adhesive or free. Adhesive eggs will stick to each other, and onto a given substrate. Free eggs can be floating, sinking, or have same specific gravity as water. When artificially incubating eggs, adherent eggs are usually cleaned and the sticky substance removed (urea, salt, tannin).

Incubation eggs need a continuous supply of oxygen, and usually in large quantities. Pure, clean, sterile water is ideal, and most hatcheries will use borehole water. Eggs should be protected from shock, as the cell divisions occurring inside the eggs can be disrupted. Direct sunlight is often a problem, and should be avoided. The eggs release carbon dioxide, which needs to be removed, and ammonia. Several kinds of egg incubators have been developed.

Causes of Egg Mortality and Treatment of Eggs

The eggs will develop normally if kept in optimum conditions. However frequently there are mortalities that are difficult to explain. This may be from a number of causes. Dissolved oxygen depletion can be a cause, a sharp variation in the temperature, incomplete fertilization, shocks and so on can be causes of mortality. As soon as the eggs die, they become white and opaque, where as viable eggs continue to show colouration and some transparency.

The presence of small crustaceans, copepods, insects or insect larvae can be a nuisance to the eggs, as they injure the shell by scratching at them. Bacteria and fungus, in particular fungus of the genus Saprolegnia, attack dead and alive eggs, and can cause the death of live eggs.

A clean water supply is therefore essential, and some filtration of the incoming water may be necessary. The best treatment against fungus is with malachite green.

A dose of 5 ppm is used for 30 to 60 minutes, or a dose of 0.1 or 0.2 ppm is introduced into the culture system and allowed to flow through. This later technique is used as a prophylaxis in very large hatcheries.

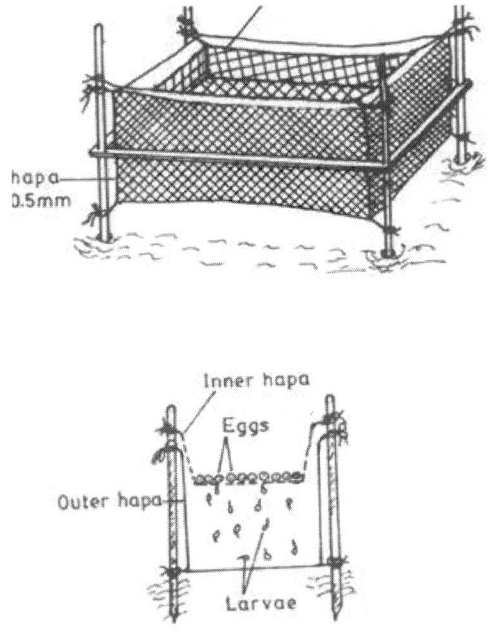

Figure: *Breeding hapa and hatching hapa used in India (After Woynarovich and Horvath, 1980)*

Formalin can also be used to treat fungus and bacteria. Concentrations of 200 to 400 ppm are used for 15 to 30 minutes. This does tend to affect the eggs and care must be taken.

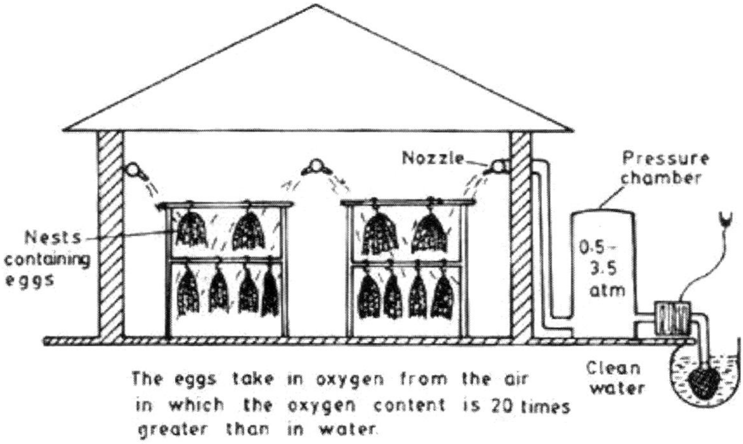

Figure: *Spray chamber for incubation of hard eggs sticking to nests (After Woynarovich and Horvath, 1980)*

Accelerating Hatching

Hatching occurs when the shell is torn, which is a mechanical and biochemical action, allowing the larva to escape. This process can be alowed or accelerated by the manipulation of temperature. If the water is too warm, the egg will hatch before the larva is well developed. If the water is cooler, the larva will be more developed than normal.

One can accelerate hatching by treating the eggs with an alkaline protease enzyme similar to the one used in laundries. The recommended doses is 0.4 to 0.5 g to 200 to 500 cc. After 3 to 5 minutes, the shells are dissolved and the larvae are free.

To synchronize hatching, one can also transfer eggs which are completely developed into water that is a few degrees warmer but with low dissolved oxygen (25% saturation). The combination of warmer water and low oxygen will cause simultaneous hatching (Bakos et al, 1975).

Larval Activity

After hatching, the larvae are more or less active. It is best to separate them as soon as possible from the debris of egg and dead eggs. If the larvae are actively swimming, there is no problem, but for some of the more inactive larvae, it is more delicate. The same mass of eggs will not hatch naturally at once, and some larva may be lost if they can not become free of the rest of the egg mass. For this reason, simultaneous hatching using enzymes can be a help. Larval activity is usually specific. Some swim to the surface, and then come back to the bottom by gravity, others will latch onto a substrate, sometimes through the secretion of a gland on the head. Others rest on the surface or bottom, occasionally displacing themselves. Finally, some form schools, and swim actively, but this usually occurs one or two days from hatching. After hatching, the larvae are still not well adapted to the environment. The digestive tube is not developed, there is no mouth, no intestine, no anus, no gills, swim bladder, scales or colouring. Feeding is on the yolk sack, a vestige from the egg. Respiration is done by diffusion, and a high oxygen concentration in the water is needed. They should be sheltered from direct sunlight. As soon as the yolk sac is resorbed, the fish begins to look for food and the whole features and behaviour of the animal changes. During this period of yolk sac resorption, the water must be very rich in dissolved oxygen, and free from any potential predators on the fish. Larvae can be maintained in jars similar to hatching jars, where a fine mesh filter retains the fish in the jar, in boxes where the sides are of fine mesh material and placed in running water, or in hapas. They can also be kept in aerated aquarium or tanks.

Organisation of a Hatchery

A hatchery can serve for the artificial propagation of one or more species. When designing the hatchery, the most important consideration is the water supply in terms of quality and quantity. Borehole water or water from a nearby spring are ideal as the temperature is constant, and the water is relatively free of organisms and pollutants. If this is not possible, a good system of filtration should be incorporated in the design of the hatchery. The physical installations should be designed to allow modifications in the future as techniques change or improve. Inflow pipes should be of a larger diametre than immediately needed, the building larger, etc. If possible tanks, incubators, troughs, and other recipients should be movable to allow for future modification. Water pipes, compressed air pipes, and other supply structures should be easily accessible, for example suspended from the roof, or placed against the walls. Probably the worst situation would be to build massive concrete structures, water and air supply pipes cemented into place. Any future modifications would be very costly and time consuming.

Broodstock Holding Facilities

Sexes should be kept separate. If using ponds, the ponds should be large enough (3 – 10 ares) to not stress the fish, should be easily and quickly drainable, with a good abundant water supply for fast refilling. As broodstock are usually very large, care must be taken to guard the fish against theft.

Hatchery and Annexes

Mature brood stock will be captured from the pond, and kept in an isolated tank, usually 1 × 2 × 1 m, for hormone treatments. Running water and/or compressed air should be supplied. The tank should be covered with a opaque or semi-opaque top, so as not to unduely stress the fish and prevent it from jumping out of the tank. Incubation materials, jars, boxes, etc. should have ready access to the hatchery workers, and should be of sufficient number to allow for all contingencies. Recipients for the newly hatched larvae should be sufficient, and readily accessable.

At all times, the hatchery designer and manager should keep in mind cleanliness and hygiene. Should the hatchery become infested with a bacteria or fungus; the manager should be able to disinfect the entire hatchery quickly and easily.

Facilities for Newly Hatched Fry

The newly hatched fry are usually kept in special containers until the yolk sac is recorded. At this point, they are transferred to either ponds or running water facilities, depending on the species and hatchery design. Ponds, or first stage fry ponds, are carefully prepared in advance to assure a good supply of natural feed for the fish, usually zooplankton. Running water facilities will usually be circular tanks, to allow close supervision of the fish. Feeding is done artificially.

Larval and Early Fry Rearing

As soon as the young fish acquire a swim bladder and the yolk sack is resorbed, they begin hunting for food. At this stage, they are very susceptible to predation. The reserves of energy in the fry are very small, and the fish must find food quickly or will die. The farmer must assure that there is abundant food of the proper quality and particle size.

Feeds that are Used and can be Used by the Larval Fish

The first natural food of the young larval fish is usually zooplankton, and among the zooplankton, the rotifers and Paramecium are usually eaten first, followed by the cladocerans and copepods. In an artificial medium, it is difficult to feed the fish with a natural feed, with the exception of the nauplii of Artemia, or brine shrimp, which can be produced by the farmer in any quantity wanted. In an artificial environment, one must replace the natural food with artificial feed. The first thing to look at is the chemical composition of the feed. The composition of natural food will vary considerably depending on their age and feeding conditions.

The average composition of Artemia salina and the rotifer Brachionus plicatilis is the following: water 85 – 95%, protein 50 – 55% (as percentage of dry matter), fats 10 – 20%, minerals 10%. As for the proteins, there is very little information on the amino acid composition, and for the great majority of species, there is very little knowledge on the essential amino acid needs, fatty acid needs, or vitamines and mineral needs. To compose an artificial feed, one can follow the rough outlines of the composition just mentioned. The size of the particles should be 0.1 to 0.3 mm in diametre. This small size makes it all but impossible to industrially pellet feed. Certain species, notably the trout and salmon, are large enough to accept larger particles of artificial feed, and these are widely used for most species, however, this is not possible.

In practical terms, artificial feeding of first-feeding larvae has several major differences than natural food. In the first instance, artificial feeds will contain only 10 – 15% water, compared to 90 – 95% water in natural feeds.

While fish larva is capable of consuming his entire weight in natural food in a single day, care must be taken when giving artificial feeds not to over feed, as this may be harmful to the newly formed digestive system and spoil the water. A second major difference is that live, natural food is usually moving, and artificial feed is inert. It is sometimes difficult to get the larva to accept artificial feed as they are not naturally attracted to it (Van der Wind, 1979).

Experience has shown that artificial feeds should contain some portion of fish meal, fish oil, or fish products. In addition to the known advantages of the amino acid composition, there is something that will attract the fish and they will accept the feed more readily if this is present. This has been shown for Clarias, Tilapia, Chrysichths, and many other species. Other feed components can include powdered baby food, powdered milk, crushed crab or fish eggs, chicken egg, and so on. The point is to attempt to have a small particle size, and a composition of roughly 50% protein.

If particle size is still too large, a simple technique, but somewhat time consuming is to place the feed mixture in a handkerchief, put the tissue in the water with the larvae, and force the mixture through the cloth with your fingers. After one or two days, the fish will usually begin to accept this feed. When artificial feeding, feeding should be several times a day, and if feasible, automatic feeders are used, the principles is to have sufficient feed available at all times. Feeding should continue until the fish have grown sufficiently to accept more conventional artificial fish feeds (agricultural by-products).

Recently, there has been some interest in the development of microencapsulation of fish feeds. Essentially, a complete feed is prepared, and encased in a shell which is biodegradable either through a lower pH as would be found in the fishes stomach, or through direct enzymes activity. A simple microencapsulated feed can be prepared by mixing an entire egg (chicken) as well as possible using a blender or mixer. Boiling water is added, and the mixture stirred. The resulting product is a microencapsulated diet with the complete protein composition of egg.

Diseases and Parasites; Prophylaxis and Treatments

A more complete description of parasites and diseases of fish and their treatment will be given in a later course. In terms of fry and larval production, there are several important differences. Even slight parasitic infections can kill larval fish, and the usual treatments for these parasites are often toxic to small fish. Great care should be taken when treating fish, and the recommended doses may be too high for the larvae.

Pond Rearing

The first nursery ponds are kept dry if they are not stocked. They are disinfected with quicklime if disease problems are prevalent, or limed with 500 kg/ha of agricultural lime if not. Water is very carefully filtered, usually the day of spawning in the hatchery, and let into the pond. The filter should be as fine as possible (0.5 to 0.8 mm). The pH optimum is between 7 and 8, and all weeds are removed. Fertilization occurs as the water is let in. Usually 5 to 7 tonnes/ha of organic manures, 150 kg/ha of triple super-phosphate, 100 kg/ha of urea. The ponds are also treated for copepods, small crustaceans and larval and adult insects. If available, a 'soft" insecticide is used, where the active ingredient is trichlorophone (Flibol L, Ditriphone, Masoten) at a concentration of 1 ppm. (Tamas & Horvat, 1979). At 17 to 20 degree the insecticide disappears in 6 to 8 days. For air breathing insects, diesel and motor oil is placed on the pond surface to stop these insects from breathing. The treatments 22 – 45 l/ha, is repeated twice a week in the absence of wind (Sneed, 1975).

The density of rotifers should be controlled before stocking the fish, the optimum density is 1500 per litre. Following the development of plankton in the pond, the relative abundance of the various groups of zooplankton will evolve. Initially rotifers are plentiful, followed by the cladocerans, and after 2 to 3 weeks, the copepods are the most numerous. The use of trichlorphon kills the cladocerans and copepods, but does not affect the rotifers, so their density can be kept at the optimum.

The density for stocking the larvae varies according to the size of the pond, the rate of fertilization, and what ever further inputs given by the farmer. Densities vary from 300 – 600 m^2 for small ponds, and 1000–3000 for larger ponds.

Natural Feeding in An Artificial Environment

Larvae and first-feeding fish can be raised in circular or rectangular tanks, aquaria, etc. The densities are usually higher, and a good supply of fresh, clean water is needed. If feeding the nauplii of Artemia, the number needed is very high. For 100,000 carp larvae, they will consume 3 to 4 litres of nauplii in the first 7 days of life. Fish should be fed every ½ to 1 hour. Another method of creating natural feed in an artificial environment is to cover the bottom of the tank with straw or hay, and fertilize the water using manures, and/or chemical fertilizers. The tank is filled with 10 to 15 cm of water, and shortly a rich fauna of paramecium and rotifers will develop. The straw is removed shortly before introducing the larvae. The water is stagnant, and each day the level of the water should be increased by 3 to 5 cm of good, oxygenated water. After 3 to 4 days, one can to begin feeding mesoplancton of 0.15 to 0.25 m in diametre, Moina, and other Cladocera. After 6 or 7 days, one can begin artificial feeding using well sifted feed particles.

The major inconvenient is the cost of construction, adequate feeding of the fry and the pollution that results with the associated diseases and lack of oxygen.

Indian Seed Sector

Seed is the basic and most critical input for sustainable agriculture. The response of all other inputs depends on quality of seeds to a large extent. It is estimated that the direct contribution of quality seed alone to the total production is about 15 – 20% depending upon the crop and it can be further raised up to 45% with efficient management of other inputs. The developments in the seed industry in India, particularly in the last 30 years, are very significant. A major re-structuring of the seed industry by Government of India through the National Seed Project Phase-I (1977-78), Phase-II (1978-79) and Phase-III (1990-1991), was carried out, which strengthened the seed infrastructure that was most needed and relevant around those times. This could be termed as a first turning point in shaping of an organized seed industry. Introduction of New Seed Development Policy (1988 – 1989) was yet another significant mile stone in the Indian Seed Industry, which transformed the very character of the seed industry. The policy gave access to Indian farmers of the best of seed and planting material available anywhere on the world. The policy stimulated appreciable investments by private individuals, Indian Corporate and MNCs in the Indian seed sector with strong R&D base for product development in each of the seed companies with more emphasis on high value hybrids of cereals and vegetables and hi-tech products such as Bt. Cotton. As a result, farmer has a wide product choice and seed industry today is set to work with a 'farmer centric' approach and is market driven. However, there is an urgent need for the State Seed Corporations also to transform themselves in tune with the industry in terms of infrastructure, technologies, approach and the management culture to be able to survive in the competitive market and to enhance their contribution in the national endeavour of increasing food production to attain food & nutritional security.

Policy Initiatives in Seed Sector

The following policy initiatives have been taken by the Government of India in seed sector:-

- Enactment of the Seeds Act, 1966
- Seed Review Team-SRT (1968)
- National Commission on Agriculture's Seed Group (1972)
- Launching of the World Bank aided National Seeds Programme (1975-85) in three phases leading to the creation of State Seeds Corporations, State Seed Certification Agencies, State Seed Testing Laboratories, Breeder Seed Programmes etc
- Seed Control Order (1983)
- Creation of the Technology Mission on Oilseeds & Pulses (TMOP) in 1986 now called The Integrated Scheme of Oilseeds, Pulses, Oil Palm and Maize (ISOPOM).
- Production and Distribution Subsidy
- Distribution of Seed Mini-kits
- Seed Transport Subsidy Scheme (1987)
- New Policy on Seed Development (1988)
- Seed Bank Scheme (2000)
- National Seeds Policy (2002)
- The Seeds Bill (2004)
- Formulation of National Seed Plan (2005)
- National Food Security Mission (2007)
- Rashtriya Krishi Vikas Yojna (2007)

  National Seeds Policy, 2002: Thrust Areas

  - Variety development
  - Plant variety protection
  - Seed production
  - Quality assurance
  - Seed distribution and marketing
  - Infrastructure facilities
  - Transgenic plant varieties
  - Import of seeds and planting materials
  - Seed exports

   - Promotion of domestic private sector seed industry
   - Strengthening of the monitoring system Seeds Bill, 2004: Salient Features
- Registration of kinds and varieties of Seeds etc.
   - Evaluation of performance
   - Compensation to Framers
   - Registration of Seed Producers and Processing Units
   - Seed dealers to be Registered
- Regulation of Sale of Seed and Seed Certification
- Seed Analysis and Seed Testing
- Export and Import of Seeds and Planting Material
- Offences and Punishment.

Seed Production System in India

The Indian seed programme largely adheres to the limited generations' system for seed multiplication in a phased manner. The system recognises three generations namely breeder, foundation and certified seeds and provides adequate safeguards for quality assurance in the seed multiplication chain to maintain the purity of the variety as it flows from the breeder to the farmer.

Breeder Seed

Breeder seed is the progeny of nucleus seed of a variety and is produced by the originating breeder or by a sponsored breeder. Breeder seed production is the mandate of the Indian Council of Agricultural Research (ICAR) and is being undertaken with the help of;

i) ICAR Research Institutions, National Research Centres and All India Coordinated Research Project of different crops;

ii) State Agricultural Universities (SAUs) with 14 centres established in different States;

iii) Sponsored breeders recognised by selected State Seed Corporations, and

iv) Non-Governmental Organizations.

ICAR also promotes sponsored breeder seed production programme through the National Seeds Corporation (NSC) / State Farms Corporation of India (SFCI), State Seeds Corporation (SSCs), Krishi Vigyan Kendras (KVKs) etc. There has been a steady increase in the production of breeder seed over the years.

The indents from various seeds producing agencies are collected by the State Departments of Agriculture and submitted to the Department of Agriculture and Cooperation (DAC), Ministry of Agriculture, Government of India, which is turn compiles the whole information crop wise and sends it to the Project Coordinator/Project Director of the respective crops in ICAR for final allocation of production responsibility to different SAUs/ICAR institutions.

The allocation of responsibility for production of breeder seed is discussed in the workshop in respect of the particular crop and is made to various centres as per the facilities and capabilities available at the centres and the availability of nucleus seed of a particular variety. It may be noted that indents are compiled and forwarded to ICAR at least 18 months in advance. To make the programme systematic, and for proper evaluation of the breeder seed production programme, monitoring terms have been constituted and reporting proformae have been devised. The monitoring terms consist of breeder of the variety, the concerned Project Director or his nominee, representative of NSC. The production of breeder seed is reviewed every year by ICAR-DAC in the annual seed review meeting.

The actual production of breeder seed by different centres is intimated to DAC by ICAR. On receipt of information from ICAR, the available breeder seed is allocated to all the indenters in an equitable manner. In the case of varieties which are relevant only to a particular State, the indents for breeder seed are placed by the concerned Director of Agriculture with the SAUs/ICAR institutions located in the State. The breeder seed produced is lifted directly by the Director of Agriculture or foundation seed producing agencies authorized by him.

Foundation Seed

Foundation seed is the progeny of breeder seed and is required to be produced from breeder seed or from foundation seed which can be clearly traced to breeder seed. The responsibility for production of foundation seed has been entrusted to the NSC, SFCI, State Seeds Corporation, State Departments of Agriculture and private seed producers, who have the necessary infrastructure facilities. Foundation seed is required to meet the standards of seed certification prescribed in the Indian Minimum Seeds Certification Standards, both at the field and laboratory testing.

Certified Seed

Certified seed is the progeny of foundation seed and must meet the standards of seed certification prescribed in the Indian Minimum Seeds Certification Standards, 1988. In case of self pollinated crops, certified seeds can also be produced from certified seeds provided it does not go beyond three generations from foundation seed stage-I.

The production and distribution of quality/certified seeds is primarily the responsibility of the State Governments. Certified seed production is organized through State Seed Corporation, Departmental Agricultural Farms, Cooperatives etc. The distribution of seeds is undertaken through a number of channels i.e. departmental outlets at block and village level, cooperatives, outlets of seed corporations, private dealers etc. The efforts of the State Governments are being supplemented by NSC and SFCI which produce varieties of national importance. NSC markets its seeds through its own marketing network and also through its dealer network. SFCI markets its seeds mainly through the State Departments of Agriculture and the State Seed Corporations. The production of certified seed by NSC and State Seed Corporations is mainly organized through contract growing arrangements with progressive farmers. SFCI undertakes seed production on its own farms. The private sector has also started to play an important role in the supply of quality seeds of vegetables and crops like hybrid maize, sorghum, Bajra, cotton, castor, sunflower, paddy etc.

The requirement of certified/quality seeds is assessed by State Governments on the basis of the area sown under different crop varieties, area covered by hybrid and self-pollinated varieties as well as the seed replacement rate achieved. The availability of seed is ascertained by the State Departments of Agriculture on the basis of the production of seed in government farms and production of seeds by State Seeds Corporations and other agencies. The Government of India periodically assesses the requirement and availability of seeds through detailed interaction with State Governments and seed producing agencies in the bi-annual Zonal Seed Review Meetings and the National Kharif and Rabi Conferences. The Department of Agriculture and Cooperation facilitates tie-up arrangements with seed producing agencies to ensure that the requirement of seeds is met to the maximum extent possible.

Information on Production and availability of Breeder, Foundation and Certified seeds can be accessed from this portal.

Role of Public & Private Seed Sector

The private sector has started to play a significant role in the seed industry over the last few years. At present, the number of companies engaged in seed production or seed trade is of the order of 400 or 500. However, the main focus of private seed companies has been on the high value low volume seeds and market for low value high volume seeds of cereals, pulses and oilseeds is still dominated by the public sector seed corporations. Private sector companies have a significant place mainly in the case of maize and sunflower and cotton. However, in the case of vegetable seeds and planting materials of horticultural crops, the private sector is the dominant player. As the private sector has not been enthusiastic about entering into seed production of high volume low margin crops of wheat, paddy, other cereals, oilseeds and pulses, the public sector seed corporations will continue to remain dominant in cereals, pulses and oilseeds for many more years to come. At present 15 State Seeds Corporation and 2 National level seeds Corporations

(National Seeds Corporation of India & State Farms Corporation of India) are functioning in the country. Besides, significant quantities of seeds are also produced by the State Departments of Agriculture, where the State Seeds corporations are not in existence. The contribution of private sector seed companies in total seed production of the country is depicted in the following table:

Year of Production	*Total Seed Production (Lakh qtls.)*	*Share of private sector*
2003-04	132.27	47.48%
2004-05	140.51	45.02%
2005-06	148.18	46.80%
2006-07	194.31	41.00%

Source: compiled by Seeds Division of DAC

Variety Registration Procedures

Each variety has to pass through 3 phases of evaluation. Breeders contribute their best entries on the basis of evaluation carried out in their local programmes for testing in the Initial Yield Evaluation Trial (IET) or Preliminary Yield Trial (PYT). These trials are organized in selected number of Places in each zone. Simultaneously, these entries are supplied to Pathologists to study their reaction to important diseases. Entries qualifying from yield, disease and quality point of view in IET/PYT are tested in the Uniform Regional Trials (URT). These trials are also called Advanced Varietal Trials (AVT) or Coordinated Varietal Trials (CVT). These trials are organized at a very large number of locations in each zone and the plot size is larger than that in IET. During the tests, reaction to various diseases, pests and quality traits are also studied. Entries found suitable in the second phase are again evaluated in the URT and simultaneously supplied to Pathologists, Entomologists, Nematologists, Agronomists and Quality Evaluation Groups to study the entries comprehensively for factors which are important from the point of view of their own discipline. Actual measurements are also made on other parameters. Agronomy group evaluates these entries for their adaptability to varied range of agronomic variables such as sowing dates, levels of fertilizers and number of irrigations etc.

These are occasionally studied for their reaction to important herbicides. These tests are followed by a critical discussion in a crop workshop. A Special Committee of multi disciplinary scientists is constituted at the workshop to consider the proposals for identification of the varieties for release. Varieties evolved by the SAUs and Government Research Institutes are tested within the concerned States at limited locations. Central Seed Committee (CSC) pointed out in 1982 that varieties of State importance might also be tested in the concerned All India Crop Improvement Project.

All the States are now submitting their samples for the coordinated trials though some have reservations of this decision. Concept of simultaneous testing of State varieties in the coordinated trials needs to be welcomed by the State Governments and their Research Institutes all over the country. Simultaneous testing of all State varieties along with the Central Varieties provides exposure to the State varieties to a wide range of environments. This will help to identify the varieties which are highly prone to diseases and pests and release of which may cause problems in some other States.

Identification of superior genotypes by Plant Breeder can benefit the public only if it is offered for commercial multiplication. It is therefore, necessary to maintain a system where quantities of promising genotype are made available for commercial production. This process is referred to as release of the varieties. The purpose of release system is to introduce newly evolved varieties to the public for general cultivation in the regions in which it is suitable. If serves as a guideline in the choice of varieties for cultivation in any region.

The practice of official release of varieties started in October, 1964 with the formation of the Central Variety Release Committee (CVRC) at the Central level and State Variety Release Committee (SARC) at State level. CVRC functioned up to November, 1969 when its functions were taken over by the CSC established Seeds Act, 1966. The CSC constituted a Central Sub-committee on Crop Standards, Notification and Release of Varieties (CSC on CS, N&RV). The sub-committee discharges the functions of release and notification of varieties at Central level, while State Seed Sub-Committees (SSSCs) discharge similar function at State level. CSC and its Sub-committee have due representation for all the agencies involved in seed research, production and quality control namely State Governments, SCAs, SAUs, ICAR Institutes, Seed producing agency in public and private sector and seed farmers.

Varietal Protection in India

India is signatory of World Trade Organization (WTO). WTO has at least half a dozen intergovernmental agreements that directly affect agriculture. Under the TRIPS Agreement Article 27(3) (b), which resulted from the negotiations of the Uruguay Round, requires members of protect plant varieties either by patents or by an effective '*sui generis*' system of protection or by a combination of both these systems. In compliance to the TRIPS Agreement India established Protection of Plant Varieties and Farmers

Rights (PPV&FR) Authority, under the Protection of Plant Varieties and Farmers Rights Act, 2001. PPV & FR Authority has become operational since 11th November, 2005. The objectives of the Authority are:

- Establishment of an effective system for protection of plant varieties, the rights of farmers and plant breeders and to encourage development of new varieties of plants.
- Recognition and protection of the rights of farmers in respect to their contribution in conserving, improving and making the available plant genetic resources for the development of new plant varieties.
- Accelerated agricultural development in the country by stimulation of investment for research and development both in public and private sector.
- Facilitate growth of seed industry to ensure the availability of quality seeds and planting material to the farmers.

Any of the following persons can make an application to the PPV & FRA for registration of a variety:-

i) Any person claimed to be a breeder of a variety.
ii) Any person being the assignee of the breeder of a variety.
iii) Any farmer or group of farmers or community of farmers claiming to be the breeder of a variety.
iv) Any University or publicly funded agricultural institution claiming to be breeder of a variety.

PPV & FR Authority shall maintain a National Register of Plant Varieties. The certificate of Registration shall be valid for a 9 years in the case of trees and vines and six years in case of other crops. The total period of variety shall not exceed 18 years for trees and vines and 15 years for extant varieties notified under Seeds Act and for other crops.

PPV & FR Authority shall invite claims for beneficiary in respect of any variety for which registration has been granted. The PPV & FR Authority shall determine beneficiary on the basis of following:-

a) The extent and nature of the use of genetic material of the claimant.
b) Commercial utility and demand in market of the variety relating to which benefit has been claimed.

The benefit determined by the PPV & FR Authority shall be deposited by the breeder with the National Gene Fund. The amount of benefit sharing shall be recoverable as arrear of land revenue. Certificate of Registration shall confer an exclusive right on the breeder, his successor, his agent or licensee the right to produce, sell, market, distribute, import or export the variety.

Farmer who has developed or bred a new variety shall be entitled for registration as a breeder of a variety. Farmer shall be deemed to be entitled to save, use, sow, re-sow, exchange, share or sell his farm produce including seed of a variety protected under this Act in the same manner as he was entitled before coming into force of this Act provided that the farmer shall not be entitled to sell branded seed of a variety protected under this Act. Farmers' variety shall be entitled for registration.

Farmer who is engaged in the conservation of genetic resources of land basis and wild relatives of economic plans and their improvement and preservation shall be entitled to recognition and reward from the Gene Fund provided the material so selected and preserved has been used as a donor of genes in varieties register able under the PPV & FR Act. Any person or group of persons (whether actively engaged in farming or not) or any other Governmental or Non-governmental organization may stake a claim on behalf of the village or local community.

There is a provision for compulsory licensing to meet the reasonable requirement of the public for seed or other propagating material. Further information can be accessed from the PPV&FR Authority's official website.

Seed Certification System in India

In general, seed certification is a process designed to maintain and make available to the general public continuous supply of high quality seeds and propagating materials of notified kinds and varieties of crops, so grown and distributed to ensure the physical identity and genetic purity. Seed certification is a legally sanctioned system for quality control of seed multiplication and production.

History of Seed Certification

The origin of the concept of seed certification dates back to the earlier part of the twentieth century. The seed certification concept grew out of the increased concern for the rapid loss of identity of varieties during production cycles. For this, credit should go to the Swedish workers who are the first to initiate the process of field evaluation of the seed crops. It began with the visits of agronomists and plant breeders to the fields of progressive farmers who took the seeds of new varieties from the. This was primarily to educate them on seed production. This initiated the process of field inspection and later on found to be very helpful in keeping varieties pure in the production chain, but other problems appeared. To overcome some of these problems, scientists from USA and Canada met in Chicago, Illinois in 1919 and formed an International Crop Improvement Association (ICIA), which later on 1969 changed its name to Association of Official Seed Certifying Agencies (AOSCA) paving the way for modern day seed certification. In India the field evaluation of the seed crop and its certification started with the establishment of National Seeds Corporation in 1963.

A legal status was given to seed certification with the enactment of first Indian Seed Act in the year 1966 and formulation of Seed Rules in 1968. The Seed Act of 1966 provided the required impetus for the establishment of official Seed Certification Agencies by the States.

Maharashtra was the first State to establish an official Seed Certifications Agency during 1970 as a part of the Department of Agriculture, whereas Karnataka was the first State to establish the Seed Certification Agency as an autonomous body during 1974. At present 22 States in the country have their own Seed Certification Agencies established under the Seed Act, 1966. In great majority of the countries in the World, including India, seed certification is voluntary and labelling is compulsory.

Objective of Seed Certification

The main objective of the Seed Certification is to ensure the acceptable standards of seed viability, vigour, purity and seed health. A well organized seed certification should help in accomplishing the following three primary objectives.

- The systematic increase of superior varieties;
- The identification of new varieties and their rapid increase under appropriate

and generally accepted names.

- Provision for continuous supply of comparable material by careful maintenance.

Eligibility Requirements for Certification

Any variety to become eligible for seed certification should meet the following requirement:

- General requirements;
- Field standards;
- Specific requirements ;
- Seed Standards;

General Requirements

- Should be a notified variety under Section-5 of the Indian Seed Act, 1966.
- Should be in the production chain and its pedigree should be traceable.

Field Standards

Field standards include the selection of site, isolation requirements, spacing, planting ratio, border rows etc.

Specific Requirements

Presence of off-types in any seed crop, pollen-shedders in Sorghum, Bajra, Sunflower etc., Shedding tassels in maize crosses, disease affected plants, objectionable weed plants etc., should be within the maximum permissible levels for certification.

Seed Certification Agencies

Seeds Act, 1966 provides for the establishment of Seed Certification Agencies in each State. Seed Certification Agency should function on the following broad principles:

- Seed Certification agency should be an autonomous body.
- Seed Certification Agency should not involve itself in the production and marketing of seeds.
- The Seed Certification Standards and procedures adapted by seed certification agency should be uniform, throughout the country.
- Seed Certification Agency should have close linkage with the technical and other related institutions.
- Its long-term objective should be to operate on no-profit no-loss basis.
- Adequate staff trained in seed certification should be maintained by the Certification Agency.
- It should have provision for creating adequate facilities for ensuring timely and through inspections.
- It should serve the interests of seed producers and farmers/users.

Organizational Set-up of a Seed Certification Agency

Organization and establishment of a Seed Certification Agency needs careful planning. In planning one should take into consideration the pertinent data about anticipated acreage for certification of various crops/varieties, area of operation, farm sizes etc. The organizational set up of the Certification Agency includes Board of Directors, technical and other staff for operating the programme. Seed Certification Agency may have its own seed testing laboratory or it may get its seed samples tested through seed testing laboratories.

Seed Certification Consists of the Following Control Measures

An Administrative check on the origin of the propagating material: Source seed verification is the first step in Seed Certification Programme. Unless the seed is from approved source and of designated class certification agency will not accept the seed field for certification, thereby ensuring the use of high quality true to type seed for sowing of seed crops.

Field Inspection : Evaluation of the growing crop in the field for varietal purity, isolation of seed crop is to prevent out-cross, physical admixtures, disease dissemination and also ensure crop condition as regards to the spread of designated diseases and the presence of objectionable weed plants etc.

Sample inspection: assessing the planting value of the seeds by laboratory tests.

Certification agency draws representative samples from the seeds produced under certification programme and subjects them to germination and other purity tests required for conforming to varietal purity.

Bulk Inspection: Under certification programme provision has been made for bulk inspection. Hence, the evaluation of the lot for the purpose of checking homogeneity of the bulk seed produced as compared with the standard sample is carried out. This gives an idea about the genuinity of lot and sample.

Control plot testing: Here the samples drawn from the source and final seed produced are grown side by side along with the standard samples of the variety in question. By comparison it can be determined whether the varietal purity and health of the produced seed are equal to the results based on field inspection.

Grow-out test: Evaluation of the seeds for their genuineness to species or varieties or seed borne infection. Here the samples drawn from the lots are grown in the field along with the standard checks. Growing plants are observed for the varietal purity. Grow-out test helps in the elimination of the sub-standard seed lots.

Phases of Seed Certification

Seed Certification is carried out in six broad phases listed as under:

i. Verification of seed source, class and other requirements of the seed used for raising the seed crop.
ii. Receipt and scrutiny of application.
iii. Inspection of the seed crop in the field to verify its conformity to the prescribed field standards.
iv. Supervision at post-harvest stages including processing and packing.
v. Drawing of samples and arranging for analysis to verify conformity to the seed standards; and
vi. Grant of certificate, issue of certification tags, labelling, sealing etc.

Seed Export / Import

The export/import of seeds and planting material is governed by the Export and Import (EXIM) Policy 2002-07 and amendment made therein. Restrictions on export of all cultivated varieties of seeds have been removed w.e.f. 01.04.2002, except the following:

(i) breeder or foundation or wild varieties;

(ii) onion, berseem, cashew, nux vomica, rubber, pepper cuttings, sandalwood, saffron, neem, forestry species and wild ornamental plants;

(iii) export of niger which is canalized through TRIFED, NAFED, etc.

(iv) groundnuts, exports of which is subject to compulsory registration of contract with APEDA;

The export of these seeds is restricted and is only allowed on case-to-case basis under licence issued by Director General Foreign Trade on the basis of the recommendations of Department of Agriculture and Cooperation.

The provisions regarding import of seeds and planting material are as under:

(a) import of seeds/tubers/bulbs/cuttings/saplings of vegetables, flowers and fruits is allowed without a licence in accordance with import permit granted under Plant Quarantine (Order), 2003 and amendment made therein.

(b) import of seeds, planting materials and living plants by ICAR, etc. is allowed without a licence in accordance with conditions specified by the Ministry of Agriculture;

(c) import of seeds/tubers of potato, garlic, fennel, coriander, cumin, etc. is allowed in accordance with import permit granted under PQ Order, 2003.

(d) import of seeds of wheat, rye, barley, oat, maize, rice, millet, jowar, bajra, ragi, other cereals, soybean, groundnut, linseed, palmnut, cotton, castor, sesamum, mustard, safflower, clover, jojoba, etc. is allowed without licence subject to the New Policy on Seed Development, 1988 and in accordance with import permit granted under PQ Order, 2003.

The EXIM Policy reiterates that all imports of seeds and planting material would be regulated under the Plant Quarantine Order 2003. Import licences would be granted by DGFT only on the recommendations of DAC. A small quantity of seeds sought to be imported would be given to ICAR, or farms accredited by ICAR, for trial and evaluation for one crop season. On receipt of applications for commercial import, DAC would consider the trial/evaluation report on the performance of the seed and their resistance to seed/soil borne diseases.

DAC is required to either reject or recommend the application to DGFT for grant of import licence within 30 days of receipt. All importers have to make available a small specified quantity of the imported seeds to the ICAR at cost price for testing/accession to the gene bank of National Bureau of Plant Genetic Resources (NBPGR). The import of seeds has to be cleared/rejected by Plant Protection Adviser (PPA) after quarantine checks within three weeks.

The rejected consignment has to be destroyed. During quarantine, the imported consignment is kept in a bonded warehouse at the cost of the importer. While importing seeds and plating material, it has to be ensured that there is absolutely no compromise on plant quarantine procedures. Every effort has to be made to prevent the entry into India of exotic pests, diseases and weeds that are detrimental to the interests of the farmers.

An EXIM Committee was constituted in the Seeds Division to deal with application for exports/imports of seeds and planting materials in accordance with the New Policy on Seed Development and EXIM Regulations. The Committee meets every month, subject to tendency of proposals for import/export of seeds and planting material, and analyses applications and furnishes recommendations to PPA/DGFT for issuing of otherwise of the licence for import/export of seeds and planting material.

Exporters/importers are required to submit 20 copies of applications for export/import in the prescribed formats. The minutes of the EXIM Committee are posted on the Seednet Portal.

As per World Seed Trade Statistics, India has sixth largest size of domestic seed market in the world, estimated to be at about 1300 million dollars. However, India's share in global trade in seeds (import & export) is of only about 37 million dollars only. To give a boost to seed export, India has decided to participate in OECD Seed Schemes for the following categories of crops:

- Grasses and legumes
- Crucifers and other oil or fibre species
- Cereals
- Maize and sorghum
- Vegetables

OECD Seed Schemes is one of the international frameworks available for certification of agricultural seeds moving in international trade. The objective of the OECD Seed Schemes is to encourage use of seeds of consistently high quality in participating countries. The Scheme authorizes use of labels and certificates for seed produced and processed for international trade according to agreed principles. The Joint Secretary (Seeds) in the Department of Agriculture & Cooperation has been nominated as the National Designated Authority.

Further, Heads of Seed Certification Agencies in Karnataka, Andhra Pradesh, Tamil Nadu, Maharashtra, Rajasthan, Uttaranchal, Uttar Pradesh Haryana, Bihar and Assam have been nominated as the Designated Authorities under the Scheme to undertake certification work under OECD Seed Schemes. The department is in the process of completing other formalities under the OECD Seed Scheme guidelines before the certification work gets started.

Conclusion

The Indian Seed Improvement Programme is backed up by a strong crop improvement programme in both the public and private sectors. At the moment, the industry is highly vibrant and energetic and is well recognised in the international seed arena. Several developing and neighbouring countries have benefited from quality seed imports from India. India's Seed Programme has a strong seed production base in terms of diverse and ideal agro-climates spread through out the country for producing high quality seeds of several tropical, temperate and sub-tropical plant varieties in enough quantities at competitive prices. Over the years, several seed crop zones have evolved with extreme levels of specialisation.

Similarly, for post harvest handing, the Indian seed processing/conditioning industry has perfected the techniques of quality up-gradation and maintenance to ensure high standards of physical condition and quality. By virtue of the diverse agro-climates several geographical zones in the country have emerged as ideal seed storage locations under ambient conditions. In terms of seed marketing and distribution, more than about 20000 seed dealers and distributors are in the business.

Over the years, seed quality specifications comparable to international standards have been evolved and are adopted by the Indian Seed Programme in both the public and private sectors. The country has a strong rigorous mechanism for seed quality control through voluntary seed certificate and compulsory labelling monitored by provincial level Seed Law Enforcement Agencies. For seed technology research, India has a national level Directorate under the Indian Council of Agricultural Research as well as Status level research set up in the State Agricultural Universities. In seed education, 4-5 prominent State Agricultural Universities offer post graduation in Seed Technology leading to M.Sc./Ph.D Degree. The seed industry has three well reputed national level associations apart from several provincial level groups to take care of the interests of the industry.

Thus, the Indian Seed Programme is now occupying a pivotal place in Indian agriculture and is well poised for continued growth in the years to come. National Seeds Corporation, which is the largest single seed organization in the country with such a wide product range, pioneered the growth and development of a sound industry in India. NSC, SFCI, States Seeds Corporations and other seed producing agencies are continuously and gradually expanding all its activities especially in terms of its product range, volume and value of seed handled, level of seed distribution to the un-reached areas, etc. Over the past four decades, these seed producing agencies have built up a hard core of competent and experienced seed producers and seed dealers in various parts of the country and have adequate level of specialisation and competence in handling and managing various segments of seed improvement on scientifically sound and commercially viable terms.

Carp Seed Production in India

Considering its significance constant efforts have been made to produce large quantity of carp seed every year in increasing trends. For instance, the total fry production in India was estimated at 632 million in 1986-87 which had increased to 18.5 billion in 2002-2003 and in 2005-06 it was over 22.6 billion. Quantified data on larger size fingerlings and/or yearlings are not available, although it is much needed for grow out culture.

Fish seed production includes egg to spawn production for 3 days, spawn to fry nursing for 15-20 days, fry to fingerling rearing for 60-90 days and fingerling to yearling rearing for 8-9 months. Thus the carp seed may be categorised at its final size into spawn (6-8 mm size), fry (20-25 mm size), fingerlings (100-150 mm size) and yearlings (100-200 g weight).

The distribution system of carp seed is complex and dynamic. Though some of the entrepreneurs produce and supply the fish seed to end users often as a part of complex networks, their supply remains erratic in other part, particularly in rural sectors1. The gap between demand and supply of quality seeds, by and large, remains a daunting task in rural aquaculture development. This can be mitigated, if village farmers produce quality carp seed in their ponds to not only make the access of locally produced and nursed quality seed to the fish farmers but also stimulate and support neighbouring farmers to adopt fish culture within their situation.

Earlier studies indicate that paucity of carp spawn compelled village farmers to stock their ponds with riverine fish seed and due to lack of technical support and basic infrastructure facilities; carp breeding was rarely adopted by farmers. In view of this various attempts have been made to demonstrate carp breeding, spawn to fry rearing and fry to fingerling rearing and fingerling to yearling rearing in rural area. Despite pointed extension focus in this regard, the sustainability aspect of the production of carp seed by the farmers still remains a missing link. Present communication summarises the carp spawn production and seed rearing management by the fish farmers of Orissa by citing examples of some selected cases.

What is Rural Carp Seed Production?

Rural carp seed production may be defined as "carp seed production by small-scale households or communities using mainly extensive and semi-intensive management appropriate to existing resource base for their own use and/ or improving their family income" or "carp seed production using technologies adapted to locally available and limited resources of households".

Rural carp seed production is not very capital intensive or input intensive and contributes to rural livelihoods. It is different from more commercially carp seed production systems or entrepreneurial carp seed production.

Evolving Rural to Entrepreneurial Carp Seed Producers

The system of carp seed production process is a continuum and it is very difficult to strictly divide rural from entrepreneurial fish seed producers. In fact, many farmers who have been involved in subsistence level carp seed production increased their production over the years, with the more inputs and better management skill, resulting in enlarging their resource base and gradually becoming entrepreneurial.

For example, a farmer who used to stock spawn in unprepared pond because of not knowing the technique of pond preparation, when came to know, followed the technique strictly and got better recovery and more income.

Thus over a period of few years he could afford more inputs and intensifies his management and becomes entrepreneurial seed producer. It is more desirable to make the resource poor farmers entrepreneurial farmers in rural area. Such evolution is already taking place with the time. For instances, farmers of Sarakana village evolved as carp seed entrepreneurs from traditional carp seed producer.

Spawn Production in Rural Area

Common Spawn Production

In rural areas generally carp spawn are generally produced twice during June-August and January-March of the year, following the adaptive breeding methods.

Pond breeding: Common carp brood fish are reared in composite fish culture ponds. In season, clean aquatic weeds such as Hydrilla / Najaj or water hyacinth are placed in pond's corners or inside floating bamboo frames in the evening hours.

During late night to early morning fish breed naturally and eggs are attached to aquatic weeds. Since water hyacinth is floating, the eggs get attached on the roots only. The egg loaded aquatic weeds are collected in morning hours and kept for incubation in hatching hapas or directly spread in well prepared nursery ponds. However, in nursery spread eggs the spawn survival is very poor than hapa hatching. This method has certain disadvantages like: difficult to estimate eggs, egg predation by pond animals, poor egg fertilisation etc.

Hapa breeding: Brood fish are reared either in separate ponds or in composite fish culture ponds. Brood fish are netted out to segregate mature males and females.

They are weighed and kept in breeding hapa containing suitable egg collectors in evening hours. Generally 3-4 kg Hydrilla/ kg female fish is used as egg collector. Males and females are kept in ratio of 1:1 by weight. They breed naturally in hapa after 6-8 hrs. In less suitable condition fishes are injected with inducing hormones to ensure breeding.

After spawning, the females are weighed to estimate the egg release. About 12-15 per cent of the weight difference goes towards faecal matter of fish and rest weight difference is due to egg release in ovary. One gram weight difference in ovary provides an estimate of 700 egg release.

Egg attached 2-4 kg Hydrilla is spread per inner hatching hapa. Depending on water temperature, hatching takes place in 2 days and inner hatching hapas are removed in 3 days. After 4-5 days, spawn are collected for stocking in nursery ponds.

Hatchery breeding: Some of the village hatchery owners use breeding pools for common carp spawning. They use nylon threads or plastic threads or plastic nets or Hydrilla or water hyacinth as egg collectors. Egg incubation is carried out in hatching pools.

Indian and Exotic Major Carp Spawn Production

Hapa breeding: In remote villages brood fish are grown in composite fish culture ponds. During monsoon season they are netted out and fully mature males and females are selected.

Breeding hapas are fixed in composite fish culture ponds having common carps. Presence of common carp, prawns and crabs cause severe damage to carp eggs in breeding hapas. Hence, to avoid hazards of loss of viable eggs, the breeding hapas are fixed inside the net enclosure. Generally for one female two males are used. Intra-muscular and/or intra-peritorial injection is administered to brood fish during June-October.

Females are injected with PG extract or glycerine extract of PG twice but males are injected only once. First dose is given in the evening hours to female @ 5-6mg/kg and second dose after 4-6 hours of first injection @ 8-16 mg/kg. Males are injected at the time of second dose of female @ 4-5mg/kg male. Presently synthetic hormones (ovaprim or ovatide) are used as inducing agents in rural areas. Both the males and females are injected only once. These synthetic hormones are administered @ 0.2-0.5 ml/ kg female and 0.1-0.2 ml/ kg male. After 4-6 hours of injection fish spawn. Fertilised eggs are identified and quantified at comma stage of embryos and hatching are done using hapa hatching device. Spawn are collected after 72-80 hours of hatching by filtering with inner hatching hapa with the spawn recovery of only 24-44 per cent of the fertilised eggs.

The low recovery of spawn from hapa hatching device could be due to a combination of factors such as cutting of hapas by crabs and/or large freshwater prawns, entry of unwanted fishes in hatching hapas8, presence of predatory cyclopoid copepods in hatching hapas and sudden change in water temperature, depletion of DO content, water bloom and cyclonic weather.

Hatchery breeding: For hatchery breeding, brood stocks are maintained in separate ponds by stoking 1-3t/ha brood fish under scientific management. Brood fishes are injected with inducing hormones as mentioned in hapa breeding. In rural areas the spawning is done in breeding hapa and/or spawning pool but hatching is done in incubation pools.

Two-three year old carps weighing 2-5 kg are the best for hypophysation. "Eco-hatchery" is used by the village entrepreneurs. It includes overhead tank, spawning pools, egg collection chamber, incubation pools and spawn collection chamber.

An overhead tank is generally made on the roof of single or double storied building and a water holding capacity of 5000 litre can supply water to spawning and incubation pools. Depending upon the requirements, the sizes of spawning pools vary. Spawning pool is 8-9 m diametre and 1.0-1.5 m deep with the provision of water circulatory system and shower.

Farmers use 20-30 kg female per spawning pool and produce 250-400 litres of carp eggs in one operation. These eggs are incubated in 3-5 hatching pools. Incubation pools are 3-4m inside diametre and 1 m deep. Generally 1 egg is incubated in one ml water.

During egg incubation, farmers maintain water flow @ 2.5 l/sec. initially, @ 2.0 l / sec at twisting movements of embryos and @ 3.5 l/sec after hatching to get better spawn recovery.

Farmers harvest 800,000 to 1,000,000 spawn/pool/operation. KVK/TTC, CIFA designed and fabricated portable FRP carp hatchery in 1989 with the maximum spawn recovery of 3,000,000 lakh / operation/pool, now modified and commercialised by CIFA and it is used by the village entrepreneurs to produce carp spawn. From hatchery breeding farmers get 80-95 per cent recovery from the viable eggs. By adopting circular carp hatchery some of the rural fish farmers changed into entrepreneurial seed producers.

Success Cases of Carp Spawn Production

Carp spawn production at Sarakana: Farmers from the Sarakana village started carp spawn production in 1987 with common carp and produced 3.5 lakh spawn in hapa - breeding. Gradually they learnt the induced breeding techniques of Indian major carps and exotic carps in hapa. Carp spawn production increased to 1,440,000-8,555,000 up to 1995.

The spawn recovery was poor and ranged between 24-44 per cent. To mitigate the problems of poor recovery of spawn in hapa, they have been motivated by KVK/TTC, CIFA to construct a cemented circular hatchery in 1995 which resulted higher spawn recovery of 74-85 per cent from 1996 onwards.

This resulted in producing 15,750,000-31,950,000 spawn of Catla catla, Labeo rohita, Cirrhinus mrigala, Cyprinus carpio, Ctenopharyngodon idella and Hypophthalmychthys molitrix annually. To meet the market demand of carp spawn in the region, they ploughed back their hard earned money to construct another carp hatchery with higher production efficiency. As a result of which they are able to produce 100-150 million carp spawn annually. This suggests that traditional seed production in rural area transformed into entrepreneurial seed production by utilising the improved technology. They produce carp seed not only to meet the market demands but also earn handsome income and employment.

Carp spawn production at Kantapada: Farmers from Kantapada village initiated carp spawn production in 1996 using hapa breeding device. With spawn recovery of 25-40 per cent of viable eggs, they produced 40, 50, 67 and 42 lakh spawn during 1996, 1997, 1998 and 1999. After realising the poor spawn recovery, farmers constructed one circular carp hatchery and now they are producing 40-60 million carp spawn annually.

Carp spawn production at Bhatapadgarh: Carp breeding was started with hapa breeding with the technical guidance of CIFA, Kausalyagang in stored rain water in ponds constructed at hilly terrains during 2002. Farmers have been trained through participatory approach in carp breeding skills. During skill learning farmers could produce 1,100,000 carp spawn with 20-40 per cent recovery in hapa breeding. Meanwhile, they constructed one commercial carp hatchery during 2003 and made it operational through the technical guidance of the author in 2004. Now farmers are producing 50-110 million spawn of C. catla, L. rohita, C. mrigala, C. carpio, L. calbasu, C. idella and H. molitrix every year at the hilly terrains to meet the regional demand.

By seeing the economic profitability in carp spawn production, many of the neighbouring farmers and entrepreneurs have constructed carp hatchery to produce carp spawn to meet the local carp seed demand.

Carp Fry and Fingerling Rearing in Rural Areas

In rural area, spawn to fry nursing is carried out in smaller ponds of 0.02-0.05 ha (0.5-1.0m depth). In same perennial ponds fry, fingerling and/or yearlings are reared in succession during June-July, August-November and December-June respectively. Alternatively the ponds are stocked with carp fry and rearing of fingerlings and yearlings are continued in succession.

For rearing larger size carp fingerlings 0.05-0.1 ha with an average depth of 1.0-2.0m are preferred. Ponds shaded by trees are rendered unproductive by reduced sunlight. Accumulation of leaf litter and an excessive organic load in the pond further deteriorates water quality, adversely affecting carp and carp food organisms. At times, masses of foamy brown/white frog eggs, which tend to fall into ponds during rains, caused a proliferation of tadpoles.

Therefore, marginal trees and bushes are cleared before launching the seed raising programme. Pond embankments are renovated with the provision of secured inlet and outlet. Since backyard ponds are shallow and small, aquatic weed clearance is completed manually by rural farmers. Predatory animals/ fishes and weed fishes are eradicated by de-watering and drying the ponds or application of suitable piscicides. Raw cattle dung is applied as basal manure in ponds. To enhance the fertilisation effect liming is done. For sustained production of natural fish food organisms a mixture of de-oiled cake, cattle dung/ bio-gas slurry and single super phosphate or a multiplex pre mineral mixture and vitamins are used in liquid forms before 4-5 days of spawn stocking.

Fry are harvested and/or thinned in phases according to the local demand, allowing an extended period of rearing (14-44 days) in rural area. Prolonged retention of fry in nursery ponds adversely affects the fry survival. Fry recovery is 20-40 per cent. Stocking spawn at shallow water depth (35-45 cm) followed by phased increase of water level at 3 - 4 days intervals, results higher fry recovery of 50-70 per cent. Fortification of micro-nutrients in artificial feeds is also enhances the growth and survival of fry. A commercially available multiplex pre-minerals mixture with vitamins accelerates plankton production and fry survival in nursery ponds. In case ponds are used for fry rearing, fry are harvested by repeated netting on day 15-20 of stocking.

At times, two crops of fry are taken. After fry harvesting, the ponds are fertilised with the mixture of above manure to produce adequate natural fish-food organisms. On day 2 or 3 of fertilisation, the fresh fry are stocked along with residual fry in such a way to maintain the density of 300,000-500,000/ ha. Later a mixture of above fertilisers is applied in liquid form at weekly or fortnightly intervals. Fingerlings are also fed traditionally and harvested by repeated netting after three months of rearing.

Success Cases of Fry and Fingerling Production

Fry and fingerling production at Sarakana village: Farmers from the Sarakana village started carp fry raising in one pond of 0.08ha and produced only 220,000 fry and 40,000 fingerlings. High profitability in fry and fingerling rearing work encouraged the farmers to invest money for creating more facilities by constructing two other ponds in 1988 and produced 384,000 fry and over 100,000 fingerlings. Since then every year the farmers expanded their activities by excavating new ponds and at preset 23 ponds of 0.02-0.1ha each are available for fry and fingerling production. Now they are producing 4,300,000-6,000,000 lakh fry and 440,000- 570,000 fingerlings every year.

Fry and fingerling production at Kantapada village: In this village fish seed nursing was initiated in 1983 by using 12 nursery ponds. Ponds were prepared and stocked @ 30-50 lakh spawn/ha. The fry were harvested after 30-45 days with the recovery of 15-30 per cent. With the time farmers acquired scientific management practices and expanded rearing area to 20 ponds (2.0 ha) gradually. Farmers are harvesting carp fry within 12-20 days with the recovery of 35-60 per cent. Multicropping of fry production is also done. They are able to harvest 3,000,000-7,600,000 fry annually. The same ponds are used for fingerling rearing with the production of over 300,000-600,000 fingerlings every year.

Fry and fingerling production at Bhatapadagarh village: Terrace type a series of 17 nursery and rearing ponds (0.05-0.17ha) are constructed with a network of inlets and outlets systems during 2003 to store huge quantity of water flowing in from the hilly terrains. These ponds were prepared by manuring, liming and insect control and stocked with carp spawn @ 3,000,000-6,000,000/ha. Ponds were harvested after 20-30 days of rearing with the recovery of 20-60 per cent yielding about 5,900,000 fry from July to September in 1 or 2 crops. After developing confidence in economic profitability, the farmers also started using even large size ponds of 0.5-0.7ha for stocking carp spawn at shallower depth followed by phased increase of water level for commercial fry and fingerling production. They are producing 6,000,000-15,000,000 fry and 100,000-800,000 fingerlings of catla, rohu, mrigal, calbasu, common carp, silver carp, and grass carp every year for supply in the region.

Large Sized Fingerling and Yearling Production

Yearlings are produced traditionally in village ponds. When farmers fail to sale their fingerlings and they continue to rear them up to May-June. Before monsoon, when ponds are prepared for next fry rearing crops, farmers harvest stunted fish for consumption as they are grown with reduced nutrient uptake. But now a days with the increased awareness of yearlings significance as stocking materials, it is being sold at pond site for grow out fish culture.

When stunted fingerlings are kept on a high quality diet they grow rapidly leading efficient body weight. Some of the village fish farmers produce yearlings and/or stunted fingerlings with improved management on commercial scale. In this, the fingerlings stocked in well prepared ponds at high density July-August. Yearlings are also reared by stocking appropriate carp fingerlings along with residual stock of fingerlings.

During culture period ponds are fertilised monthly once. Fingerlings are fed with the mixture of ground nut oil cake and rice bran in the ratio of 1:1 by weight @ 4-6 per cent of the body weight. Complete harvesting of yearlings is done by repeated netting from May-June. Adopting this management the farmers of Kantapada and Bhatpadagarh are producing 3-5 tonnes of yearlings every year.

Crop Management

Planting Materials

Seed

Seed is a fertile and ripened ovule, which contains an embryonic plant, supplied with food storage tissues and surrounded by a seed coat. A seed strictly speaking is an 'embryo' a living organism embedded in the supporting or the food storage tissue. According to the Seed Act 1966, seed includes all planting materials such as seedlings, tubers, bulbs, grafts, etc.

Seed Technology - involves seed production, processing, storage testing and certification, marketing and distribution and the related research on these aspects. Seed production includes varietals development, evaluation and release of varieties.

Role of Seed Technology - improved seed - a carrier of new technologies a basic tool for secured food supply- the principal means to secure crop yields in less favourable areas of productivity - a medium for rapid rehabilitation of agriculture in areas of natural disaster - a 'national seed reserve stock' of seed is necessary to face such situations.

Seed is the fundamental unit of crop production - Regarded as the most vital, basic and critical input in agriculture for increasing and sustaining agricultural production. The year in which the Royal Commission on Agriculture was constituted -1925. It emphasized the need for a separate organization in India to deal with seed distribution. This paved the way for the set up of the National Seed Corporation (NSC) in 1963. Coverage of high yielding varieties in Kerala -60.32% (1999-2000). Maximum coverage during Punja followed by Viruppu and Mundakan. The poor coverage is the non-availability of quality seeds of the High Yielding Varieties (HYV). In India seed production and distribution have become a well developed industry - involving an annual turn over of Rs.2200 crores - Public sector -Rs.55 crores (25%) - Private sector Rs.1320 crores (60%) - other unorganised sector Rs.330 crores (15%) India is exporting seeds worth of Rs.20 crores in an year. Goals - To increase agricultural production through the spread of good quality seeds of HYVs. It aims at (1) rapid multiplication, (2) timely supply, (3) assuming high quality, (4) reasonable price.

Bases of a Seed Programme - (1) high level support, (2) productive plant breeding programme, (3) Co-ordinated effort.

History of development of Seed Technology

- First Five Year Plan (1951-56) - Distribution of improved seeds of food grains came into existence - seed distribution with subsidy.
- Second Five Year Plan (1956-61) - Improved seed was made as a basis for 10% additional food grain production - Production of foundation seeds - started seed production centres and seed stores.
- Third Five Year Plan (1961-66) - Released first four hybrids of maize in 1961 - necessitated a separate organization for seed production. Thus gave birth to NSC in 1963.

HYV was started during 1966 and there was high coverage of HYV seeds for food crops by 1968-69. The HYV Programme had an almost explosive effect on the seed industries and necessitated huge effort in seed production, processing, certification, testing and storage, seed distribution, etc.

Annual Plans (1966-69) - Seed Review Team - Constituted by Government of India in 1968- suggested to entrust the work related to production, processing and marketing of seed co-operative and private sectors - also NSC should assist State Government to constitute their own certification agencies. IV plan (1969-74) - In 1969 - Tara Development Corporation Ltd was started with the active involvement of GB Panl University of agriculture and technology - Growers participation as shareholders instead of contract system of seed production.

Seed Act 1966 Came into Force through out India from 1969: V Five Year Plan (1974-79) - Reviewed by National Commission on Agriculture - Many recommendations were made in 1976 - (1) Rigorous enforcement of Seed Act 1966; (2) National Registry of varieties - system by ICAR; (3) Departments of Agriculture - in the centre and state should have three distinct wings each dealing respectively with (a) input aspect, (b) land enforcement (c) certification; (4) Grow-out tests should be made on integral parts of seed testing.

Generation Concept of Seed Multiplication

NUCLEUS SEED - initial seed-small quantity from the originating plant breeder - for further multiplication to maintain the variety - multiplication by the supervision of the plant breeder or any other qualified plant breeder

- genetic purity - 100%. BREEDER SEED - progeny of nucleus seed multiplied by the breeder or by the institution - source for the foundation seed - genetic purity is 100%. Not subjected to certification. But monitoring by a team of exports - Golden Yellow colour tag.

FOUNDATION SEED (FS) - Progeny of Breeder seed production is supervised by certification agency. Minimum genetic purity 99%-white tag. Can be II stages. FS Stage FS Stage II and I. Stage II is only if Breeder Seed I in short supply. NSC, SDA and State Seed Corporation are involved in the production.

CERTIFIED SEED (CS) - Produced from FS, Seed Certification Agency certifies - supervising by seed certification agency, State Seed Corporation and NSC - subjected to one or more stage of production genuine shortage of FS. Such production does not exceed three generations beyond FS.1 Minimum genetic purity 98% - Tag colour Azure Blue. To maintaining genetic purity, Cs should be replaced every 4th season or 4th crop.

6

Seed Certification

Purpose: To maintain and make available high quality seeds with genetic identify and purity - designed to achieve prescribed standards - legally sanctioned system for quality control during all stages of seed production. In India seed certification is a voluntary process - but labelling is compulsory.

NSC evolved the certification standards. The central seed committee approved these standards - the standards are includes in the Indian minimum seed certification standards published by Government to establish their own seed certification agencies.

A statutory central seed certification Board co-ordinated the activities of State seed certification agencies. (Only those varieties, which are notified by central seed committee, are eligible for certification).

Seed Certification Phases

- Verification of seed source
- Verification and approval of seed fields prior to planting
- Inspection of seed crop in field to verify its conformity to the prescribed standards.
- Supervision of harvest and post harvest operations including processing and packing.
- Sampling and analysis of seeds to conform seed standards.
- Grant of certificate tags, lebelling and sealing.

Seed Certification standards

			(Percentage)
Quantity Traits	ES	CS	TLS
Pure seed (minimum)	98	98	97
Other crop seeds (maximum)	10 No./Kg	0.1	-
Weed seed (maximum)	10 No./Kg	0.1	-
Inert matter (maximum)	2	2	-
Objectionable weed seed (maximum)	2 No./Kg	5 No./Kg	-
Germination (minimum)	80	80	70
Moisture content	12	12	-

Source: Indian minimum seed certification standards 1971 - Central Seed Committee)

Truthfully Labelled Seeds (TLS): Seeds of the varieties, which are not eligible for certification can be, produced as truthfully labelled seeds. The seeds should be labelled as follows:

(1) Kind, (2) variety, (3) Lot No, (4) Date of test, (5) Inert matter % (6) pure seed %, (7) Other crop seed %, (8) weed seed %, (9) Germination %, (10) net content, (11) sellers name and address, (12) if treated, any one of the following statements should appear in the label:

'Do not use for food, feed or oil purpose (or Poison)'. If the content of the container is less than 250gm or less, the items 5 to 9 may be replaced by the statement " The seed in this container conform to the minimum limits of germination and purity prescribed under the Act"

Vegetative Propagation

Leaf Cutting

Many plants can be artificially cloned by leaf cuttings; species that work well include African violet (Saintpaulia), Peperomia, bryophyllum (Kalanchoe) and jade plant (Crassula). In this process a leaf (blade + petiole) is removed from the donor plant. The leaf is placed in moist soil or, perhaps, in water and placed in the light.

The leaf blade, as usual, produces IAA (auxin). This auxin is transported basipetally (down the petiole) as usual.

The IAA accumulates in the base of the petiole to the point that a callus tissue (tumor) forms. A callus is a mixture of meristematic cells and parenchyma cells. Since these cells do not have a determined fate, it is possible to influence these cells to differentiate in just about any direction. The IAA continues to accumulate in the callus. The concentration rises to the point that roots are initiated on the callus.

The roots produce cytokinins (CK). This hormone is transported acropetally (away from the root tip, toward the callus). The cytokinin accumulates in the callus.

As the concentration of cytokinin accumulates in the callus, it stimulates shoot formation. The shoots then can grow up. In the end you have a shoot with roots...a whole plant!

In some plants, the callus sends up multiple shoots. In those cases, the gardener has to separate the individual plantlets before potting up the individual plants.

Obviously this process can be accelerated if the last centimetre of the petiole is dusted with a mixture of talcum powder and synthetic auxin (indole butyric acid) before the leaf is planted in the moist soil at the beginning. This mixture is often called "rooting hormone" and two leading brands of this are Rootone&174 and Hormodin&174.

Stem Cutting

Many species cannot be propagated from leaves alone. The leaf may root, but shoots do not form, or form so slowly that you do not dare wait for it. For such species, it may be possible to instead make a stem cutting. The process is very much the same, but is much faster due to the contributions of auxin from several leaves, and the contributions of photosynthesis in several leaves.

Generally the gardener cuts a short piece of stem (three or four internodes maximum). It does not have to include an apical meristem, but often it does. The leaves from the lowest node are torn from the stem. The basal end of the cutting may be dusted with rooting hormone (if needed or desired).

The several leaves remaining on the stem cutting all send IAA (auxin) down from the blade. This accumulates at the basal end of the cutting (along with any applied rooting hormone). The accumulated auxin stimulates formation of callus and ultimately roots.

Once we have a stem with roots, our plant has been cloned and the process is complete.

Simple Layer

A tip layer is a natural method of propagation in such species as black raspberry. A simple layer is fundamentally an artificially-forced tip layer. In many shrubs that cannot be cloned by leaf or stem cuttings, a simple layer may work well. Rather than making a single leaf or a small stem cutting provide for its own needs and enough extra to make a new plant, a simple layer can be used to allow contributions from the whole donor plant to drive the cloning process.

1. A stem, low on the shrub, is arched mechanically to the ground and held in place with some small wire hoops.
2. The lower surface of the held stem is wounded (with a small notch or some grooves) and the wound is dusted with rooting hormone.
3. The wounded, dusted, section of the stem is buried in the moist soil.
4. With time, the hormone powder and contributions of IAA from the rest of the leaves on the donor plant cause roots to form from the wounded area.
5. The rooted stem can then be separated from the donor plant.

Air Layer

If a gardener is trying to propagate a tree instead of a shrub, there is no way to bring a branch down to the ground for a tip layer. If you cannot bring Mohammed to the Mountain, then you bring the Mountain to Mohammed! This is the idea of the air layer.

1. Somewhere along a young stem in the tree canopy, the lower surface of a stem is wounded and the wound dusted with rooting hormone.
2. A few hundred cubic centimetres of moist sphagnum moss or even moist soil is packed around the wound.
3. The sphagnum/soil is held in place with plastic film, and the film is secured around the stem with twist-ties above and below the wound. The soil has to be kept continually moist.
4. With time, roots develop in the small wad of sphagnum. The rooted branch can then be cut from the donor tree.

Tissue Culture

When one is trying to make many thousands of individuals of one clone, then a different method is needed. Tissue culture to the rescue!

1. First you need some donor tissue. This is called the explant. The tissue has to be at least parenchyma, but is better if it is meristematic. The tissue has to be surface-sterilized. This is done by immersing the tissue in a solution of sodium hypochlorite (laundry bleach), rinsing in sterile distilled water, then perhaps slicing off the killed surface layers.
2. The explant is placed in a sterilized liquid medium on a shaking machine. This medium contains water, minerals, vitamins, sugar (energy for growth by respiration), and dilute auxin (usually indole-butyric acid). The auxin stimulates the cells of the explant to divide rapidly to form callus. The callus is broken up by the shaking action. Thus the liquid medium becomes loaded with small callus pieces.

3. The callus pieces are now spread across a sterile solid medium. The medium contains water, minerals, vitamins, sugar, and cytokinins. The medium is solidified into a gel by adding a small amount of agar. The container is kept under light. The combination of cytokinin and light causes shoots to be initiated on each callus piece.
4. The shoots are moved onto a solid medium that contains water, minerals, and agar. Depending on the species, there may also be some auxin in the medium. The auxin from the leaves on the shoots (or in combination with auxin from the medium) cause root initiation.
5. The plants can be moved to soil and slowly eased from humid conditions to successively drier conditions. Ultimately the plants can be moved to field conditions.

Because each explant can initiate indefinitely many callus pieces, and each callus piece can make a new plant, the number of plants produced can be extremely high.

Fertilizers

Chemical Fertilisers

Choice of a fertiliser depends on unit cost of nutrient present in it and its agronomic efficiency under a given situation. Fertiliser is a valuable input and measures should be taken to reduce its losses and to increase its uptake and utilisation by the crop. Selecting a situation-specific fertilizer and choosing the time and method of application according to crop demand would minimise losses and increase its efficiency.

Nitrogenous Fertilizers

Most crop plants recover only 25-35% of the nitrogen applied as fertilizers. Losses occur by ammonia volatilisation, denitrification, immobilization to organic forms, leaching and run off. Utmost care should be bestowed in selecting the type of fertiliser as well as the timing and method of application.

Choice of the Nitrogen Fertilizer

1. In submerged rice soil, ammoniacal and ammonia-producing fertilizers like urea are most suitable since ammonia is the most stable form of nitrogen under such conditions.
2. For acidic upland soils, ammoniacal fertilizers are most suitable during rainy season since ammonium is adsorbed on soil particles and hence leaching losses are reduced. Adsorbed ammonium is gradually released for nitrification and thus becomes available to crops for a longer period.
3. In highly acidic upland soils, urea is preferred to ammonium sulphate as the former is less acid forming.

4. In alkaline upland soils of low rainfall regions, nitrate fertilizers are preferred to ammoniacal fertilisers or urea since ammonia may be lost by volatilization under alkaline conditions.

Management of Nitrogenous Fertilizers

1. Almost all the nitrogenous fertilizers are highly amenable to losses and since most of the crops require nitrogen during the entire growth period, split application is necessary to ensure maximum utilisation by crops.
2. More number of splits may be given for long duration crops as well as perennial crops.
3. Nitrogen losses from fertilisers are more in coarse textured soils with low cation exchange capacity (CEC) than in fine textured soils. Hence more number of splits is necessary to reduce loss of fertilizer nitrogen from sandy and other light soils.
4. For medium duration rice varieties, nitrogenous fertilizers should be given in three splits, as basal, at maximum tillering and at panicle initiation stage.
5. In coarse textured sandy or loamy soils, the entire dose of nitrogenous fertilizers may be applied in 3-4 splits at different stages of growth of rice crop.
6. In areas where split application of nitrogen is not feasible due to water stagnation after planting/sowing, full dose of nitrogen as basal may be given in the form of neem coated or coal tar coated urea.
7. In double-cropped wetlands, 50% of N requirement of the first crop may be applied in the organic form.
8. As far as possible, liming should be done one or two weeks prior to the application of ammoniacal or ammonia forming fertilizer like urea since ammonia is likely to be lost by volatilization if applied along with lime
9. Almost 70% of N in urea applied by broadcast to flooded soil is lost by volatilization, immobilization and by denitrification

Measures to Reduce the Loss of N from Applied Urea

1. Urea super granules or urea briquettes may be used in places where soil is clayey and has cation exchange capacity more than 10 cmol (+) per kg of soil.
2. Sulphur or lac coated urea is suitable where soil is liable to intermittent flooding and in situations where water management is difficult. This is more suitable for direct sown crop.

3. Urea may be mixed with moist soil and kept for 24-48 hours before application to the field. Alternatively, urea may be mixed with moist soil, made into balls of about three-inch diametre and dried under shade. The balls may be placed deep into subsoil.
4. Mixing urea with five times its weight of neem cake prolongs the period of nitrogen availability to the crop.
5. For submerged soils, coating urea with coal tar and kerosene (100 kg urea is mixed with 2 kg coal tar dissolved in one litre kerosene) before mixing with neem cake is preferred to simple mixing with neem cake.
6. Coating urea with neem extract (containing about 5% neem triterpenes) at 1% rate and shade-drying for 1 to 1.5 hours before applying in direct-seeded puddled lowland rice increases nitrogen use efficiency.
7. As far as possible, urea may be applied by deep placement or plough sole placement. Deep placement of prilled urea or super granules during the last ploughing followed by flooding and planting is beneficial in light soils. Urea briquettes or super granules may be placed between four hills of transplanted rice, whereas sulphur coated or lac coated urea may be broadcast on the surface.
8. Foliar spray of 5% urea solution can be practised in situations where quick response to applied nitrogen is required. If power sprayers are used, the concentration may be increased to 15%. Fresh urea should be used to avoid toxicity due to biuret.

Phosphatic Fertilizers

Fertilizer phosphorus is an expensive input and its management poses serious problems due to several complexities in its behaviour in different types of soil. This often results in its poor recovery from applied fertilizers.

Choice of Phosphatic Fertilizer

1. In slightly acid, neutral or mildly alkaline soils, water-soluble phosphatic fertilizers are more suitable.
2. In wetland rice soils, water-soluble phosphatic fertilizers are preferable as pH of most of the submerged soils is near neutral.
3. In strongly acidic soils whose pH does not rise above 5.5 to 6.0 even on submergence, phosphatic fertilizers containing citrate soluble form of P like basic slag, dicalcium phosphate, steamed bone meal etc. are suitable.
4. For highly acidic upland soils or submerged soils whose pH will not rise above 5.5 even on submergence, powdered rock phosphate is suitable. Soil acidity converts tricalcium phosphate in rock phosphate to plant available monocalcium form.

5. For short duration crops where quick response is required, water-soluble phosphatic fertilizers are most suitable.
6. For perennial crops like rubber, oil palm, coffee, tea, cardamom etc. phosphorus in the form of rock phosphate can be applied.
7. In black soil (Chittur taluk of Palakkad District) phosphatic fertilizers containing water-soluble phosphate like single superphosphate are most suitable.

Management of Phosphatic Fertilizers

1. Acid soils have to be amended with lime, dolomite or magnesium silicate and alkali soils with iron pyrite or sulphur before application of phosphatic fertilizers. This will help to reduce fixation and increase availability of P.
2. Surface application or broadcasting is preferred for shallow rooted crops whereas placement in the root zone is advantageous in deep-rooted crops.
3. Rock phosphates can be used advantageously in rice grown in acid soils during the virippu season. Powdered rock phosphate may be applied and mixed thoroughly with soil by ploughing. After two or three weeks, the field may be flooded, worked up and planted with rice. Under this situation, phosphorus in rock phosphate gets converted to iron phosphate, which on subsequent waterlogging becomes available to the rice crop.
4. Rock phosphate can be used successfully as a phosphatic source for leguminous crop since its root system can extract phosphorous from rock phosphate.
5. In single crop wetlands where rice is grown in the virippu season, application of phosphatic fertilizers can be dispensed with for the rice crop, if the second crop (usually legume or green manure) is given phosphatic fertilizers.
6. In case of rice-legume cropping sequence in acid soils, application of rock phosphate to the pulse crop helps to skip phosphatic fertilizers in the succeeding rice crop.
7. Since phosphorus requirement of seasonal crops is confined to the early stages, phosphatic fertilizers are to be applied at the time of seeding or planting. Topdressing of phosphatic fertilizer leads to wastage of the fertilizer nutrient. Further, excessive phosphates may lead to deficiency of micronutrients such as zinc, boron etc.
8. Under adverse soil conditions and where quick result is required, spraying water-soluble phosphatic fertilizers like triple superphosphate or hot water extract of superphosphate can be resorted to.

Potassium Fertilizers

For most crops, potassium can be supplied as muriate of potash. But in crops like tobacco and potato, muriate of potash may cause chloride injury, reducing quality of the produce. In such cases, K may be applied as potassium sulphate.

Management of Potassium Fertilizer

1. In coarse textured soils and in heavy rainfall regions, potassium fertilizers should be applied in as many splits as possible, to reduce loss of potassium.
2. In fine textured soils, the entire dose of potassium fertilizers may be applied as basal.
3. In acid soils, potassium fertilizers should be applied only after lime application to prevent loss of potassium by leaching.

Organic Manures

Organic manures are waste or by products (animal/birds waste, litter, crop refuse or any other organic matter) either decomposed or treated or fresh, which enriches soil. They are bulky i.e. nutrient contents are low per unit volume. Concentrated OM contains higher % of nutrients such us oil cakes, blood/bone/fish meal important effects are:

1. improve the physical properties of the soil
2. supply all nutrients for plant growth
3. enhances soil microbial activity
4. provide a buffering action in soil reactions
5. improve nutrient holding capacity of the soil

Green Manures

Green manuring refers to the practice of incorporating plant materials while they are green. It is practiced as green leaf manuring and green manuring insitu.

Average Nutrient Content of Common Organic Manures

Major Nutrients in Organic Manures			
	N	*P2O5*	*K2O*
Neem cake	5.2	1	1.4
Coconut cake	3	1.9	1.8
Groundnut cake	7	1.5	1.3
Gingelly cake	6.2	2	1.2

Castor cake	4.3	2	1.3
Mahua cake	2.5	0.8	1.8
Farm yard manure	0.5 - 1.5	0.4 - 0.8	0.5 - 1.9
Compost (Rural)	0.4 - 0.8	0.3 - 0.6	0.7 - 1.0
Compost (Urban)	1.0 - 2.0	1	1.5
Cowdung (fresh)	1.5	0.3	0.2
Biogas slurry	—	—	—
(cowdung based)	1.6 - 1.8	1.1 - 2.0	0.8 - 1.2
Sheep manure	0.8 - 1.6	0.4 - 0.6	0.3 - 0.4
Poultry manure	1.2 - 1.8	1.4 - 1.8	0.8 - 0.9
Bone meal	3.5	21	—
Fish meal	4.1	3.9	0.3
Meat meal	11	1.5	0.6
Blood (dried)	11.5	—	0.6
Sun hemp	2.6	0.5	2
Daincha	3.3	0.7	1.3
Kolinji	3.2	0.3	1.3
Sesbania	2.7	0.5	2.2
Gliricidia	2.9	0.5	2.8
Paddy straw	0.36	0.08	0.71
Household ash	0.5 - 1.9	1.6 - 4.2	2.3 - 12.00
Coir pith Compost	1.26	0.06	1.2

Note: Composition of organic manures vary widely

Bio-fertilizers

The use of biofertilizers is quite important while practicing the concepts of integrated plant nutrient management and organic farming. Some of the commonly used biofertilisers in Kerala are as follows.

1. *Rhizobium (Bradyrhizobium and Azorhizobium):* It induces better root nodulation and stem nodulation (Azorhizobium) in inoculated plants and thereby brings down the requirement of nitrogen fertilizer for the cultivation of pulses, oil seeds and legume green manures. Commercially it is available as carrier based inoculums. Method of application is seed treatment.
2. *Azotobacter:* Suitable only for upland crops like vegetables, tapioca, plantation and orchard crops. It is available as carrier-based inoculum. It fixes N about 15-20 kg/ha under ideal upland conditions and thereby

reduces the requirement of nitrogen fertilizers by 10-20 per cent. Methods of application are seed treatment, seedling dip and direct soil application.

3. *Azospirillum:* It is suitable for both upland and wetland conditions and is available as carrier-based inoculum. It fixes N about 20-25 kg per ha under ideal conditions thereby effecting a reduction of 25 per cent in the quantity of N fertilizers required. Treatment with Azospirillum also induces better root formation in inoculated plants. Hence this biofertilizer is also recommended for root induction in polybag-raised seedlings of plantation and orchard crops and also for vegetable crops. The isolates of Azospirillum brasilense strains AZR 15 and AZR 37 from Kuttanad soils are highly effective for rice, vegetables and nursery plants. The strains AZ 1 and AZ 2 are effective in vegetable and nursery plants.

Method of Application

Seed treatment: For treating 5-10 kg seeds, 500 g culture is required. Moisten the seeds by sprinkling water or rice-gruel water. Take 500 g culture in a plastic tray/basin, add moistened seeds, mix well and dry in shade for 30 minutes. This may be sown immediately.

Seedling root dip (for transplanted crops): Slurry of the culture is prepared by mixing 500 g culture with 50 ml of water and the roots are dipped in the slurry for 15-20 minutes before transplanting. Soil application: Mix the culture with FYM or compost in the ratio 1:25 and apply directly in the soil.

Inoculation for Paddy

Mix 2 kg of culture in 60 litres of water and soak the seeds required for 1 ha (60 kg) for 24 hours before sowing. At the time of transplanting, dip the roots of seedlings for 15-20 minutes in the culture slurry prepared by mixing 2 kg inoculum with 40 litres of water. This slurry can be used for treating seedlings required for 1 ha. Another 2 kg culture may be applied to the field along with FYM or compost.

Blue Green Algae (BGA): Mainly recommended for wetland rice cultivation. However, the use of this biofertilizer is not feasible in acidic soils with pH below 6.0. It is available as carrier-based inoculum and it fixes N about 25-30 kg/ha under ideal conditions.

Method of application: Direct broadcasting in the rice fields @ 10 kg/ha one week after transplanting the seedlings.

Azolla: It is suitable for wetland rice cultivation. The required quantity of azolla will have to be raised in the farmers' field itself. Fixes N about 25 to 30 kg / ha. Method of application: Apply fresh azolla @ 10 t/ha before transplanting the rice seedlings at the time of ploughing.

Phosphate solubilising bacteria and fungi: Recommended mainly for upland crops raised in neutral and slightly alkaline soils. Available as carrier-based inoculum. Enables the efficient utilisation of cheaper sources of phosphatic fertilizers such as rock phosphate by crop plants in neutral and alkaline soils. Method of application: Seed treatment and direct application.

Vesicular / Arbuscular Mycorrhiza (VAM/ AM): Vesicular arbuscular mycorrhiza is mostly recommended for upland especially for raising container and tissue culture plantlets and transplanted crops. It mainly improves the uptake of available P by inoculated plants.

There is also an enhanced absorption of water and other nutrients such as N and K and certain micronutrients. Mycorrhiza inoculation can improve the survival and establishment of tissue culture plantlets under field conditions. Also induces better resistance against certain soil borne plant pathogens. It is commercially available as granular inoculum consisting of infected roots and soil with mycorrhizal spores. It is given as soil application.

VAM fungus Glomus microcarpum var. microcarpum is suitable for tropical tuber crops. The inoculation can be done by placing inoculum (3-5 g/sett) beneath the sett before planting. The rate of spore load in the inoculum should be to the tune of 50 to 400 spores per 100 g soil medium. Method of application is the rooted infected cutting technique.

Application Techniques of Biofertilisers

1. *Seed treatment:* Five hundred grams of commercially available inoculum will be required for treatment of seeds for one-hectare area. For this, thick slurry of the carrier-based inoculum is initially prepared by mixing 500 g of the inoculum in 1.25 litre of water. The stickiness of the biofertilizer on seed surface can be significantly improved by using 10% jagerry solution or 5% sugar solution supplemented with 40% boiled and cooled gum arabic solution or rice-gruel water. The required quantity of seed material is then gently mixed with this slurry by taking care not to damage the seed coat. The treated seeds are spread evenly over a gunny bag and dried in shade and sown immediately in moist soil. Under no circumstances, the treated seeds are exposed to direct sunlight for a longer period of time since the UV rays of solar radiation will reduce significantly the population of inoculated bacteria on seed surface.
2. *Seedling treatment:* This method of application is mainly recommended for transplanted crops. In this procedure, the roots of seedlings to be transplanted are dipped in loose water slurry of the biofertilizer (500 g in 2.5 litre of water) for 20 minutes, prior to transplanting.

3. *Soil application:* Soil application is generally recommended for all types of biofertilisers except Rhizobium, Bradyrhizobium and Azorhizobium. The method is to apply the biofertilizer after mixing with dried FYM, compost or vermicompost @ 1:25. For crops of six-month duration, the recommended dose is 1-2 kg/ha. This can be increased to 2-4 kg/ha for crops of more than six-month duration. For perennial crops, 10 to 25 g of the biofertilizer is to be applied in the root zone during the first year and 25 to 50 g during subsequent years. This can be done at the time of sowing, transplanting or during intercultivation.

Factors Influencing the Efficient use of Bio-Fertilisers in Kerala

1. Use adequate quantity of organic manure (as per the recommendation for each crop) along with biofertiliser application. This is essential to ensure better survival, growth and activity of the introduced microbial inoculums in acidic soils.
2. Liming is essential if the soil pH is below 6.0. In moderately acidic soils, the application of lime @ 250 kg/ha is recommended along with biofertilizer treatment.
3. Irrigation is essential during summer months after biofertilizer application to ensure the survival of the introduced microbial inoculum in the soil.
4. Since N biofertilisers can supplement only a part of the nitrogen requirement of the inoculated plant, low dose of nitrogen and full doses of phosphorus and potassium as per the recommendation may be applied. This is essential to ensure better plant growth and yield. Similarly, in the case of P biofertilisers, the full doses of nitrogen and potassium should be applied. However, there should be a gap of at least one week between biofertilizer and chemical fertilizer applications.
5. Use only biofertilisers, which are manufactured as per the quality parameters prescribed by the Bureau of Indian Standards. In the case of bacterial biofertilisers, the prescribed standard is that in the final product, the population of the desired bacterium should not be less than ten million per gram of the carrier material and there should not be any contamination with other microorganisms when examined at 1:100000 dilution. Further, it should have a shelf life of at least six months.
6. The commercially available biofertiliser should always be used before the expiry date marked on the culture packet.
7. Topdressing with super phosphate 25 kg/ha 10 days after inoculation of BGA will enhance its growth under field conditions.

8. Since the occurrence of green algae in rice field can affect the normal growth and proliferation of BGA, the population of green algae should be controlled initially by applying copper sulphate @ 4 kg/ha.
9. In moderately acidic soils of pH around 6.5, root nodulation by Rhizobium and Bradyrhizobium can be improved by pelleting with finely powdered calcium carbonate.
10. Application of P2O5 @ 1 kg/ha is recommended once in 4 days in P2O5 deficient soils to ensure good growth of azolla. The development of a reddish purple colour in azolla is a typical symptom of P2O5 deficiency.
11. Since a floating population of azolla can release its bound nutrients only during decay in the soil, it is essential to incorporate azolla in the soil prior to the transplanting of rice seedlings.

Liming

Lime

Acid soils are characterised by high saturation of the exchange complex with hydrogen and aluminium. Crops grown in such soils suffer due to unavailability of most plant nutrients, especially calcium. Application of liming materials increases the availability of nutrients and alleviates Ca deficiency.

Liming Materials

Burnt lime [CaO], slaked lime [Ca(OH)2], powdered limestone [CaCO3] and dolomite [CaMg(CO3)2] are some of the materials used as sources of calcium.

Management

1. In acidic submerged soils, flooding brings about rise in soil pH and hence response to lime is less marked.
2. Legumes are benefited most by liming.
3. For better results, liming materials should be incorporated into the soil.
4. For seasonal crops and in situations where immediate results are required, burnt lime or slaked lime may be used. For perennial crops, powdered limestone or dolomite is sufficient.
5. Extreme care should be taken while broadcasting burnt lime and slaked lime as they can cause scorching of leaves.
6. In case of wetland rice, drain the field prior to lime application and re-flood after 24 hours. Flushing the soil by sequential flooding and draining will help to wash out the displaced acid from the soil.
7. In extreme case of calcium deficiency, a 1% solution of calcium chloride may be applied by foliar spraying.

Plant Protection

Cultural

Cultural Control:

> *"Cultural control is the deliberate manipulation of environment to make it less favourable for pests by disrupting their reproductive cycles, eliminating their food, or making it more favourable for their natural enemies"*

Cultural control is an important component of Integrated Pest Management (IPM). It is one of the oldest and most effective methods of pest suppression. Cultural control methods are widely used in pest management systems. These methods comprise of regular gardening operations, the purpose is to destroy pests or to prevent them from causing injury to gardens. Examples Include cultivated techniques, adjustment of planting dates, choice of cultivars (resistant varieties), general crop hygiene, manipulation of harvesting procedures, crop rotation, crop covers, etc. Below are descriptions of some of these methods to help you choose ways to help deter pests in your garden.

Variety Selection:

Choose resistant plant varieties.

Planting Dates:

Crop Rotation:

Crop rotation can be very important in fighting disease and pests. Rotate crops year to year. Do not plant the same things in the same place year after year

Trapping Crops: Some pests prefer certain plants to others. By planting sacrificial plants, you can keep away pests from the plants you want.

Row Covers: A floating row cover is a physical barrier used to protect crops, particularly annual vegetables. Floating row covers should be used when the damage is being done by the insect larvae. It prevents the adult flying insect from landing and depositing eggs on the plants. The floating row cover should be placed over the entire crop at transplanting or seeding time and it must cover the plants.

Chemical

Pesticide application has become more complex over the past several years. The number of different kinds of pesticides available for use has increased. Effects on wildlife and the environment are now known to be important considerations. Highly toxic pesticides require special equipment and safety measures. Pesticide applicators, therefore, need to know more about safety and proper use than ever before. New applicators must know to protect the environment from pesticide contamination, the proper handling of pesticides, and protection from personal injury.

Types of Pesticides

A pesticide is any chemical which is used by man to control pests. The pests may be insects, plant diseases, fungi, weeds, nematodes, snails, slugs, etc. Therefore, insecticides, fungicides, herbicides, etc., are all types of pesticides. Some pesticides must only contact (touch) the pest to be deadly. Others must be swallowed to be effective. The way that each pesticide attacks a pest suggests the best way to apply it; to reach and expose all the pests. For example, a pesticide may be more effective and less costly as a bait, rather than as a surface spray.

Insecticides

Insecticides are chemicals used to control insects. Often the word "insecticide" is confused with the word "pesticide." It is, however, just one of many types of pesticides. An insecticide may kill the insect by touching it or it may have to be swallowed to be effective. Some insecticides kill both by touch and by swallowing. Insecticides called Systemics may be absorbed, injected, or fed into the plant or animal to be protected. When the insect feeds on this plant or animal, it ingests the systemic chemical and is killed.

Broad Spectrum. Insecticides vary in the numbers of different kinds of insects they kill. Some insecticides kill only a few kinds of insects. Sometimes you can choose these insecticides when you wish to kill only one insect pest and not other beneficial insects in the area. Many insecticides are general purpose or wide range killers. These "broad spectrum" pesticides are used when several different kinds of insects are a problem. One chemical can kill them all. No broad spectrum insecticide kills all insects; each varies as to the kinds of insects it controls.

Narrow Spectrum. While many insecticides are broad spectrum, killing a wide variety of animals by attacking a system common to all, such as the nervous system, a new group of insecticides are much more selective. The chitin inhibitors only affect animals with chitin in their exoskeleton (i.e. insects). Growth regulators are even more specific. They affect certain groups of species that have a particular hormone. Finally, pheromones are the most restrictive because they react with only one species or one sex of a single species. Chitin synthesis inhibitors interfere with the development and molting of immature insects causing their death. Chitin is the primary structural chemical in an insects body wall. An immature insect treated with a chitin inhibitor dies the next time it attempts to molt.

Insect growth regulators or IGRs mimic the action of an insect's naturally occurring juvenile hormone. They interfere with certain normal processes and prevent immature insects from completing development into normal reproductive adults.

The effects of IGRs on insects include abnormal molting, twisted wings, loss of mating behaviour, and sometimes death to embryos in eggs. IGRs attack a growth process found only in insects, thus there is a great margin of safety for humans and other vertebrates. However, one disadvantage is that growth regulators act slowly, since they do not kill the insect until it molts into an adult.

Pheromones

Pheromones are naturally produced chemicals used by animals to communicate to each other. There are three basic types of pheromones. Aggregation pheromones attract many individuals together, for example, a site where food may be plentiful. Sex pheromones are used by one sex of a species to attract a mate. Trail pheromones are deposited by walking insects, such as ants, so that others can follow. Synthetic pheromones produced in laboratories mimic these natural chemicals. They are used to attract pest insects into traps, disrupt mating, and monitor populations of insects. Because they do not kill insects, they are often not considered to be pesticides.

Short Term vs. Residual. Insecticides also vary in how long they last as a killing agent. Some break down almost immediately into nontoxic by - products. These "short term" chemicals are very good in situations where the insects do not return or where long-term exposure could injure non-target plants or animals. For example, short-term insecticides are often used in homes and dwellings where people and domestic animals might be exposed. Other insecticides remain active killers for a fairly long period of time. These "residual" pesticides are very useful when the insects are a constant problem and where they will not be an environmental and/or health hazard. For example, residuals are often used for fly control in livestock buildings or for termite control in wooden structures.

Miticides and Acaricides

Miticides (or Acaricides) are chemicals used to control mites (tiny Insecticides spider-like animals) and ticks. The chemicals usually must contact the mites or ticks to be effective. These animals are so numerous and small, that great care must be used to completely cover the area on which the mites live. Miticides are very similar in action to insecticides and often the same pesticide kills both insects and mites. The terms "broad spectrum," "short term," and "residual" are also used.

Fungicides

Fungicides are chemicals used to control the fungi which cause molds, rots, and plant diseases. All fungicides work by coming in contact with the fungus, because fungi do not "swallow" in the normal sense. Therefore, most fungicides are applied over a large surface area to try to directly hit every fungus. Some fungicides may be systemic in that the plant to be protected may be fed or injected with the chemical. The chemical then moves throughout the plant, killing the fungi to describe miticides.

Protectant vs. Eradicant. There are two basic approaches in the use offungicides. One is designed to prevent the plant from getting the disease. These fungicides are used as "protectants" and are similar in purpose to polio and smallpox vaccinations for humans.

They are applied before the disease gets a start. This type of fungicide is very useful when a particular disease or group of diseases are likely to attack a plant or crop, year after year. Protectants, for example, have often been used as a routine precaution on fruit and vegetable crops.

Most protectant fungicides are fungistatic. This means they prevent or inhibit fungal growth. Once the fungistatic action ceases, the controlled fungus may grow again or produce spores. Thus, a protectant fungicide may have to be applied at regular intervals to continue the protection from infection.

The other type of fungicide kills the disease after it appears on (or in) the plant. These fungicides, called "eradicants," are like penicillin or other antibiotics which cure diseases in humans after the sickness appears. Eradicants are less common than protectants because once the fungus is established in a plant, it is often difficult to destroy.

Eradicants are often used when protectants aren't available, aren't applied in time, or are too expensive. Eradicants are also applied when the disease appears unexpectedly on a plant or in an area. For example, a common use is on fruit and vegetables when the protectant spray wasn't applied on time to prevent infection. Eradicants are also used by orchardists in combatting diseases of fruit trees, such as apple scab.

Herbicides

Herbicides are chemicals used to control unwanted plants. These chemicals are a bit different from other pesticides because they are used to kill or slow the growth of some plants, rather than to protect them. Some herbicides kill every plant they contact, while others kill only certain plants. Nonselective herbicides are toxic to all plants. These are often used when no plants are wanted in an area. For example, nonselective herbicides could be used for clearing under guardrails or for total control of weeds in industrial areas.

Selective herbicides kill some plants with little or no injury to other plants. Usually selective types will kill either broadleaved plants or grassy plants. These are useful for lawns, golf courses or in areas with desirable trees. Some very selective herbicides may kill only certain plants in a group; for example, crabgrass killers on lawns.

Preplanting vs. Preemergence vs. Postemergence. The timing of an herbicide application is important. Care must be used to get the job done effectively without injuring desirable plants. The directions on the label tell you when to apply the herbicide for best results. Preplanting treatments are made before the crop is planted. These chemicals may be used in seed beds or incorporated into the soil before planting. Any treatment made before the crop and weed appears is called preemergence. The application may be made before both the crop and weeds appear, or after the crop appears but before the weeds appear. The label or directions will state "preemergence to the crop," "preemergence to the weeds," or "preemergence to both crop and weeds."

When the herbicide treatment is made after the crop or weeds appear, it is called postemergence. Postemergence applications must be very selective. They must control the weeds but leave the crop unharmed. Often, the chemical will be applied postemergent to the crop but preemergent to the weeds.

Growth Regulators and Harvest Aids

A plant growth regulator (or plant regulator) increases, decreases or changes normal growth or reproduction in a plant. Fertilizers and other nutrients are not included. Some growth regulators are used to move up or move back the normal harvest date for the crop. Others are used to obtain better quality and/or yield of the crop. Electric power utilities could use growth regulators to slow the growth of a tree threatening power lines, thus saving the tree from being cut.

Defoliants and desiccants are pesticide materials generally referred to as harvest aids. A defoliant causes the leaves of a plant to drop off early, but does not kill the plant. A desiccant draws moisture from a plant, killing the plant foliage.

Rodenticides

Rodenticides are chemicals used to control rats, mice, bats and other rodents. Chemicals which control other mammals, birds, and fish are also grouped in this category by regulatory agencies. Most rodenticides are stomach poisons and are often applied as baits. Even rodenticides which act by contacting the pest are usually not applied over large surfaces because of the hazard to domestic animals or desirable wildlife. They are usually applied in limited areas such as runways, known feeding places, or as baits.

Nematicides Molluscicides Repellents

Nematicides are chemicals used to control nematodes. Nematodes are tiny hir-like worms, many of which live in the soil and feed on plant roots. Very few of these worms live above ground. Usually, soil fumigants are used to control nematodes in the soil. However, a few contact insecticides and fungicides are also effective against these tiny worms.

Molluscicides are chemicals used to control snails and slugs. Usually the chemicals must be eaten by the pest to work. Baits are often used to attract and kill snails or slugs in an area.

A repellent is a pesticide that makes a site or food unattractive to a target pest. They are registered in the same way other pesticides are and must be used according to the label.

Insect repellents are available as aerosols and lotions and can be applied to skin, clothing, or plants to repel biting and nuisance insects. Vertebrate repellents are available as concentrates to be mixed with water, powders, and granules. They can be sprayed or painted on nursery crops, ornamental plantings, orchards, vineyards, vegetables, and seeds. Repelling deer, dogs, birds, raccoons, and others can protect sites from damage.

Handling and Mixing

Take precautions to avoid exposure when wettable powders, dust or granules are added to the sprayer tank. When you add materials to the sprayer tank, air is forced out and carries some of the pesticide particles with it. If the solvent in the pesticide is toxic or flammable (or both), be sure the mixing operation is performed in an area where ventilation is adequate. The addition of small amounts of materials such as emulsifiers or thickeners will drastically alter the physical properties of the spray solution. Therefore, the applicator should check the pesticide label to be sure she or he is operating according to instructions.

Many types of solvents, some of which are chlorinated, are used in the pesticide formulation processes. Vapors of chlorinated solvents are very dangerous to breathe. They can cause a "high," dizziness, or even unconsciousness. They also can cause permanent damage to the kidney, liver, and nervous system in workers exposed to the vapours for a prolonged time.

Field Applications

Always be aware of the meteorological conditions existing during pesticide application. Loss of spray from a treated area increases during high winds or low humidity. Usually, low wind and high humidity conditions are most prevalent before 10 a.m. and after 6 p.m.

Avoid spraying near beehives, lakes, streams, pastures, houses, schools, playgrounds, hospitals or sensitive crops whenever possible. If these areas must be sprayed, do not spray during windy or low humidity conditions and always spray downwind from the sensitive area.

If possible, begin spraying in the end of the field that will permit any drift from the sprayer to be blown away from the next area to be treated. Likewise with airblast sprayers; direct the air blast from the sprayer with the prevailing wind and away from the next area to be treated. These two procedures will minimise the amount of pesticide that will be blown onto the operator. Remember, these are only guidelines and are intended to supplement the good judgment of the pesticide operator.

To minimise drift hazards, use the lowest pressure possible, the lowest boom height and the largest spray tips, and add thickeners (if the pesticide label permits) in areas where drift is likely to be particularly hazardous.

Be alert for nozzle clogging and changes in nozzle patterns. If nozzles clog or other troubles occur in the field, shut the sprayer off and move to an unsprayed area before dismounting from the sprayer to work on it. If nozzles must be cleaned in the field, use a toothbrush or a toothpick for cleaning - never a metal object. A metal object can damage the orifice in the spray tip and significantly alter the spray pattern. We recommend carrying extra spray tips on the sprayer so that plugged tips can be replaced with clean ones. Never try to unclog a nozzle by blowing through it.

Check the pesticide label for re-entry and preharvest intervals. The re-entry interval is the elapsed time after a pesticide application before workers can safely re-enter a field. The preharvest interval is the elapsed time between a pesticide application and harvest of the crop.

Equipment Cleaning and Storage

Trained personnel should thoroughly clean the inside and outside of mixing, loading, and application equipment immediately after use. People who clean contaminated equipment should wear proper protective clothing, including rubber boots, a rubber apron, goggles, and possibly a respirator. A specific area should be designated for cleaning operations. Use a rack or cement apron with a well-designed sump to catch contaminated wash water and pesticides. The cleanup process is important because many chemicals will rapidly corrode some metals and may also react with succeeding chemicals, thus possibly causing a loss of effectiveness.

To clean a sprayer, mix about two pounds of detergent per 40 gallons of water in the tank. Circulate this mixture throughout the bypass or agitator nozzles for 30 minutes and then drain, allowing some solution to pass through the booms or nozzles. If phenoxy herbicides such as 2,4-D, 2,4,5-T, or brush killer were used in the sprayer, rinse with an ammonia or charcoal solution. To make an ammonia rinse, fill the tank one-third to one-half full and add two quarts of household ammonia per 25 gallons of water.

Circulate the solution and allow a small amount to flow through the nozzles. Allow the remaining solution to stand overnight to neutralize any herbicide remaining in the equipment. Then pump the solution through the nozzles. After rinsing with detergent or ammonia, flush thoroughly with clean water.

When the sprayer is to be stored for a prolonged time, add one to five gallons of lightweight oil (about one gallon of oil per 40 gallons of water) before the final flushing. As the water is pumped from the sprayer, the oil will leave a protective coating on the inside of the tank, pump and plumbing.

To prevent corrosion, remove nozzle tips and screens and store them in a can of light oil, such as diesel fuel or kerosene. Follow directions in the owner's manual regarding the proper procedures for storing engine-equipped sprayers. Before storing the sprayer, all lines, hoses, valves and the pump should be inspected for damaged parts or leaks. Damaged parts should be replaced before the sprayer is stored.

Storage of Chemicals

Store pesticides in a locked and posted facility where children or other untrained people cannot get to them. Read the labels on pesticide containers for correct storage procedures. Select a storage site high enough that damage from flooding will be unlikely and downwind and downhill from sensitive areas such as houses, play areas and ponds. Pesticide storage facilities should also be located away from homes and livestock facilities to avoid or minimise contamination.

Pesticide and expended pesticide containers are best stored in a separate building, room, or enclosure, depending on the size of the pesticide inventory. The storage area should be on the first floor to minimise contamination from a possible leaky container, and the storage area must keep the pesticides dry, cool and out of direct sunlight.

The latter requirements are necessary because some pesticides are rendered useless if they become too hot or damp, if they freeze, or if they are exposed to ultraviolet radiation from the sun.

Do not store pesticides near food for human consumption, animal feed, fertilizer, seed, veterinary supplies, or other stored products. To prevent contamination or to avoid accidentally using the wrong pesticide, store different pesticides in separate locations within the storage area.

Store pesticides only in the original container, with the label plainly visible. Never store pesticides in anything used as a food or drink container, even for a short time. Storage of pesticides in such containers is a common cause of accidental poisoning.

Preparing the Sprayer

Before a spraying operation is started, rinse out the sprayer; remove and clean all nozzles, nozzle screens and strainers. Make sure strainers and nozzle screens are 50 mesh or larger when wettable powders are used. Check all lines, valves, seals and the tank after filling the sprayer with water and during calibration to be sure there are no leaks in the spray system. For the operator's safety, replace weather-cracked or worn hoses. Adjust the nozzle height and spacing as suggested by the nozzle manufacturer or as specified on the pesticide label. These requirements differ for a given pest or crop.

Only water that is clean enough should be used in the sprayer. A small amount of silt or sand in the water rapidly wears pumps and other parts of the sprayer system.

Disposal of Pesticides and Containers

To avoid having to dispose of a tank load of the wrong pesticide, check out the job carefully before selecting the pesticide. After you have selected the proper pesticide, mix only enough for the particular job. Preventing a pesticide surplus is the best way to prevent a disposal problem.

Despite your best efforts, however, you cannot always avoid surplus pesticides, and you must take steps to dispose of them properly.

If you mix too much pesticide for a job, try to find other areas with the same pest problem and use any extra tank mix or rinse water on these areas. In some cases, small amounts of surplus pesticide can be diluted and reapplied to the treated area. Take extreme care to prevent excessive residues, especially with herbicides, by making sure that the total application rate does not exceed the maximum rate for which the pesticide is labelled.

To dispose of large quantities of pesticide, contact nearest krishibhawan for assistance in properly disposing of excess pesticides in an environmentally safe manner.

So-called empty pesticide containers are not really empty. They still contain small amounts of pesticides, even after they have been properly rinsed. All containers, regardless of their type, should be rinsed three times before disposal. The rinse water should be dumped into the sprayer tank. Otherwise, the rinse water must be treated as a surplus pesticide and disposed of properly. Rinse water should never be dumped on the ground. Use the following rinse-and-drain procedure to prepare containers for disposal:

Empty the container into the spray tank and drain in a vertical position for 30 seconds. Refill the container one-fifth to one-fourth full with rinse water.

Rinse thoroughly, pour into the spray tank and drain in a vertical position for 30 seconds. Repeat steps 2 and 3 until the container has been rinsed three times. Empty the container into the spray tank and drain in a vertical position for 30 seconds.

Rinsed containers should not be used for any other purpose. Disposal of any pesticide container or pesticide-related waste by open dumping or open burning is illegal.

Bio Control

The biological control of pests involves using natural enemies of the pest to control it - instead of chemical agents like insecticides and herbicides. Not only should this be safer for the environment, but - once established - the natural enemies might be able to sustain their population avoiding the need for future treatments. Most of the species that we consider pests are plants ("weeds") or animals (especially insects) that have invaded a new habitat without being accompanied by the natural enemies that kept them in check in their original home. With increasing international travel and trade, this problem becomes increasingly severe.

IPM

Integrated Pest Management (IPM) is a process consisting of the balanced use of cultural, biological, and chemical procedures that are environmentally compatible, economically feasible, and socially acceptable to reduce pest populations to tolerable levels. Integrated means that many strategies are used to avoid or solve a pest problem. These strategies come from different disciplines, such as disease information from plant pathologists, weed information from agronomists, and insect information from entomologists.

Pests are unwanted organisms that are a nuisance to man or domestic animals, and can cause injury to humans, animals, plants, structures, and possessions. Management is the process of making decisions in a systematic way to keep pests from reaching intolerable levels. Small populations of pests can often be tolerated; total eradication is often not necessary.

Reasons for Having a Broader Approach to Integrated Pest Management (IPM) Han Just the Use of Chemicals

Keep a Balanced Ecosystem. Every ecosystem, made up of living things and their non-living environment, has a balance; the actions of one creature in the ecosystem usually affect other, different organisms. The introduction of chemicals into the ecosystem can change this balance, destroying certain species and allowing other species (sometimes pests themselves) to dominate. Beneficial insects such as the ladybird beetle and lacewing larvae, both of which consume pests, can be killed by pesticides, leaving few natural mechanisms of pest control.

Pesticides Can be Ineffective. Chemical pesticides are not always effective. Pests can become resistant to pesticides. In fact, some 600 cases of pests developing pesticide resistance have been documented to date, including common lamb's-quarter, house flies, the Colorado potato beetle, the Indian meal moth, Norway rats, and the greenhouse whitefly. Furthermore, pests may survive in some situations where the chemical does not reach pests, is washed off, is applied at an improper rate, or is applied at an improper life stage of the pest.

IPM Is Not Difficult. Although some of the terms and ideas may be new to you, practicing IPM is not difficult. Believe it or not, you will have done much of the "work" for an IPM approach if you've figured out the problem (the pest), determined the extent of the damage, and decided on the action to take. These steps are the same ones used in IPM. Save Money. IPM can save money through avoiding crop loss (due to pests), and avoiding unnecessary pesticide expense. For example, farmers in Kuttanad have saved more than half there expense for plant protection after the inception of ORP for pest surveillance and forecasting.

Promote a Healthy Environment. We have much to learn about the persistence of chemicals in the environment, and their effect on living creatures. However, more cases of contaminated groundwater appear each year, and disposal of containers and unused pesticides still pose challenges for applicators. Even though long-term documentation on the effects of all pesticides is still unavailable, it is generally agreed that fewer pesticides means less risk to surface water and groundwater, and less hazard to wildlife and humans.

Maintain a Good Public Image. Recent public outcry about the use of growth regulators and the presence of pesticide residues on produce has heightened pesticide applicator awareness of the level of public concern about chemicals. Consumers are pressuring food stores, which in turn are pressuring producers, for produce that has been grown with as few pesticides as possible. Growing food under integrated pest management can help allay public concerns. Structural pest control professionals can suggest improvements in housekeeping or structural modifications as substitutes for chemical control.

The Basic Steps of IPM

All of the components of an IPM approach can be grouped into four major steps. The first step is taking preventative measures to prevent pest buildup, the second is monitoring, the third step is assessing the pest situation, and the fourth is determining the best action to take.

Preventative Measures

Many IPM practices are used before a pest problem develops to prevent or stall the buildup of pests.

Cultural Controls are those that disrupt the environment of the pest. Plowing, crop rotation, removal of infected plant material, sanitation of greenhouse equipment, and effective manure management are all cultural practices that are employed to deprive pests of a comfortable habitat.

The management of urban and industrial pests has improved when sanitation programs have been improved, pest harborages eliminated, garbage pickup frequency increased, or when lights are installed that do not attract insects.

Structural Modifications - by preventing support timbers from soil contact, damage from several different wood destroying pests can be avoided. Wood absorbs moisture and is more susceptible to attack by carpenter ants and termites when in direct contact with the soil.

Construction Site Sanitation - removing tree stumps and lumber scraps from construction sites, which are prime food sources for subterranean termites, can prevent problems in the future.

Biological Controls - using natural enemies (biological control agents) to keep pests in check can be put into place before pest problems increase.

Examples of biological control agents are beneficial mites that feed on mite pests in orchards, the milky spore disease that kills harmful soil grubs, and Encarsia formosa , a wasp that parasitises the greenhouse whitefly. Many biological control agents are commercially available.

Physical Barriers such as netting over small fruits and screening in greenhouses can prevent crop loss. Physical barriers are important in termite, housefly, and rodent control.

Use of Pheromones (natural insect scents) has become widely used in pest management. Sometimes a manufactured "copy" of the pheromone that a female insect emits to attract mates can be used to confuse males and prevent mating. This technique is used in curbing damage from the grape berry moth.

Pest-Resistant Varieties are those that are less susceptible than other varieties to certain insects and diseases.

Use of resistant varieties often means that growers do not need to apply as many pesticides as with susceptible varieties. Potato growers control the golden nematode by planting resistant cultivars. Apple growers can save up to eight fungicide applications a year by growing Liberty and Freedom cultivars, which resist diseases. Farmers growing alfalfa and wheat keep several pests at bay by planting resistant varieties. Once a pest manager has taken precautions to prevent pest infestations, it is important to watch regularly for the appearance of insects, weeds, diseases, and other pests.

Monitoring (Scouting)

Monitoring pests involves:

regular checking of the area;

early detection of pests;

proper identification of pests;

identification of the effects of biological control agents.

Regular checking of a warehouse, bakery, restaurant, field, greenhouse, golf course, or other areas and early detection of pests can function together like an early warning system for pests, helping to avoid or prevent a pest problem.

Proper identification of pests is an extremely important prerequisite to handling problems effectively. For example, the brown banded and German cockroach can be easily confused with each other. Identification is important because certain management practices may control only one species and not the other.

Correct identification enables you to manage the real source of the problem and avoid merely treating the symptoms (or controlling non-pests). Some pests cause similar evidence. Unless the pest is identified, the control program may have the wrong pest as its target. Identification enables you to cure the pest problem and avoid injury to non-target organisms, particularly if you:

Use a pesticide that is specific to the pest, control the pest effectively during the most susceptible stage of its life cycle, consider the use of a non-chemical control.

Identifying the effects of biological control means knowing which creatures are helpful and determining if pests have been affected by the beneficial organisms. Sometimes pests are kept in check naturally, and at other times the pest populations increase sharply.

Assessment is the process of determining the potential for pest populations to reach an economic threshold or an intolerable level. Is a grower likely to suffer financially? Is the pest likely to transmit a disease? How can you tell? There are important differences between the assessment of crop pests and urban pests.

Forecasting can help you determine if weather conditions will be favourable for the development of diseases and insect pests. For example, by "plugging in" values (such as the number of rainy days and the temperatures for those days), growers can predict outbreaks and spray only when conditions are favourable for diseases. Growers who have kept good records of pests in previous years can use these records to help determine if problems such as weeds, insects, and diseases will reoccur. They might be able, for example, to apply the most effective herbicides at the proper time for early control of a problem.

Thresholds, or more specifically economic thresholds, are levels that mark the highest point a pest population can reach without risk of economic loss.

Populations above these thresholds can reach the economic injury level, where they cause enough damage for the grower to lose money. At the economic injury level, the cost of control is equal to the loss of yield or quality that would result otherwise.

Thresholds for many pests and crops have been scientifically determined. The advantage of thresholds is that if a pest has not reached threshold, there is no risk of economic loss.

Therefore, there is no need to spray. Once the pest density (number of pests per unit area) has reached threshold, action is justified.

The costs of control will be less than equal to the estimated losses that the pests would cause if justify uncontrolled.

Urban pest thresholds are often related to aesthetics rather than economic considerations. Where health concerns or individual sensitivities exist, the tolerable level of the pest may be zero.

A zero threshold forces action, even if only one pest has been detected. Zero thresholds exist in hospitals, food production, warehousing, and retail facilities.

Action (Control Measures)

Once a pest has reached the economic threshold, or intolerable level, action should be taken. In some situations, cultural controls can destroy pests. One example is early harvesting to avoid pest problems, which prevents crop loss and can sometimes be more economical than a pesticide application.

Chemical pesticides are used as a control measure when no other strategies will bring the pest population under the threshold. In fact, the success of waiting until a pest reaches threshold usually hinges on the availability of a pesticide that will bring the pest populations down quickly.

In summary, an IPM approach means that pest managers use multiple tactics to prevent pest buildups, monitor pest populations, assess the damage, and make informed management decisions, keeping in mind that pesticides should be used judiciously.

Weed Management

Integrated pest management of weeds, like insect IPM, focuses on prevention, beginning with identification of weed species and controlling the population in the field.

Localized use of weed control tools may be necessary in areas with high concentrations of perennial weeds

Prevention

Avoid weed establishment; eliminate individual survivors.

Employ sanitary procedures; prevent weed spread.

Clean equipment between sites or infestations.

Examine nursery plants, seed, and imported soil or media.

Screen irrigation water where weed seed contaminates surface water transported in canals and rivers or stored in lakes or ponds.

Control weeds and seed sources around the field or site.

Establish national and state weed laws and noxious weed control programs.

Biological

Control weeds using organisms which feed/infect/pest on the weed.

Cultural

Integrate numerous components to minimise impact of weeds. Select manageable fields (identify weeds and choose crop according to feasibility of weed management strategies)

Rotate crops (disrupt weed life cycles or suppress weeds in competitive crop followed by planting a noncompetitive crop). Plant competitive fallow crops in rotation to improve soils and crop management. Consider legumes to supplement nitrogen requirements.

Consider specific vegetation which remain uniform on soil surface; either perennial or large-seeded crops can be planted through undisturbed mulch.

Alter planting dates (plant for maximum growth or delay planting to control first weed flush).

Transplant slow-growing crops. Place and time of fertilizer application, especially nitrogen is highly decisive. Band or spot fertilizer beside plant or seed (reduces availability to surface-germinating weeds). Time additional top dressings for maximum crop growth or to minimise weeds.

Develop crop canopy that shades weeds, suppresses weed germination.

Select crops or varieties that form canopy quickly.

Space plants in equidistant (triangular) arrangements and vary density depending on crop management constraints or harvest requirements.

Interplant crops in space and time.

Combine broadleaf and taller, narrow leaf crops.

Relay plantings or harvest short-duration crops within longer maturing crops

Manage appropriate living mulch (grass or legume) between perennial crop rows.

Improve pasture management by reseeding and/or fertilizing with or without control measures to reduce weed infestation (weeds often are a sign of poor management).

Apply mulch or geo-textiles

Solarisation Thin plastic secured tightly over loose, moist soil for 10 to 12 weeks will suppress weed infestations and other pests during hot season; data unavailable for Oregon.

Mechanical

Weed debris turned under; annual weeds controlled; perennial weeds suppressed if repeated every 10 to 14 days.

Tillage Practices: Seedbed preparation/planting (plough down/mix debris). Use Plough/ rotovater/ discharrow (compacts wet soils).

Cultivators for weed control. Sweeps, rolling cultivators, finger weeders, push hoe, rotary hoes, etc.

Hand pulling or hand hoeing.

Minimum or no-till systems (often requires specialised equipment).

Seedbed preparation/planting accomplished in one operation where trash is separated, soil loosened, and crop planted.

Weed control achieved with herbicides or sometimes cultivators, depending on debris.

Flaming (requires temporal difference between crop and weed growth).

Water management (drip placement or timing to reduce weed growth).

Stale seedbed (planting after controlling first weed flush).

Herbicidal

Requires precision calculations, equipment, and application.

Selectivity

Crop naturally tolerates herbicide (internal selectivity).

Placement of herbicide prevents crop exposure.

Timing of application to avoid susceptible stage of growth.

Labelling requires extensive testing.

Toxicology includes acute, subacute, and chronic toxicity; teratology (fetus); reproduction, mutagenicity (cancer); neurotoxicity (nerves); and metabolism studies along with worker exposure.

Environmental fate includes breakdown, groundwater, ecological effects on plants and animals.

Tolerance assessment considers normal application procedures and preharvest intervals for assessing residues within food products.

7

Seed Cleaning

What is Seed Cleaning?

This is the removal of any debris or low quality, infested or infected seeds and seeds of different species (weeds) that are foreign to the sample.

Why is Seed Cleaning Important?

The cost of maintaining an accession in a genebank is high and space is limited. Debris and damaged seeds can spread infection.

Therefore only good quality viable seeds should occupy space in the store and damaged or non-viable seeds should be destroyed to prevent the spread of infection. Each accession should be represented in the genebank by a minimum number of clean seeds. For Information on Sample Size, Seesection vii.

When should Seed Cleaning be Carried Out?

Seeds should be cleaned immediately after registration or harvest. If the seeds are very moist and cleaning is done by machines, it may be necessary to dry the seeds within the fruits to within 12 - 16% moisture content before cleaning to prevent damage. Seeds donated by other institutes often arrive already cleaned.

How should Cleaning be Done?

Cleaning should be done in a way that causes the least damage to the sample and does not waste good seeds. The following procedures can be used as a guide.

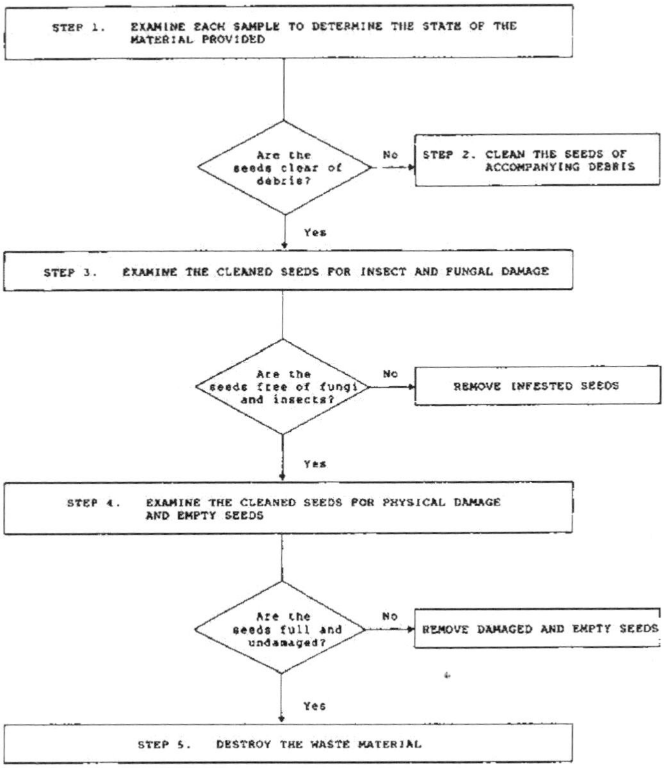

Examine Each Sample to Determine the State of the Material Provided

1. Treat each accession separately.
2. Empty the contents of each packet into a labelled tray large enough for the material to be spread out.
3. Check whether the seeds are in dry fruits or fleshy fruits or have already been cleaned.

Clean the Seeds of Accompanying Debris

Seeds from dry fruits:

1. Check that the seeds are dry enough to be cleaned without damage because moist seeds (above 16% moisture content) can be damaged during cleaning.
2. If the samples are very moist, place the container with the fruits in the drying room, if available, or in a shady place to dry.
3. When it is predicted that the seed moisture content is in the range of 12 - 16%, the seeds can be safely cleaned.

4. Clean the seeds from the fruits, preferably by hand.
5. Small threshing machines can be used for some crops, provided that they do not damage the seeds. Clean threshing machines between each accession to prevent mixing of seeds from different accessions. Mechanical cleaning is of less value in multi-crop genebanks.
6. Remove debris from the seeds by the use of gradated sieves.
7. Pieces of debris of a similar weight and size as the seeds will have to be removed by hand.
8. Separate dry lightweight material from the seeds by winnowing, blowing gently on the seeds or by use of a seed blower.
9. Return the cleaned seeds to the tray ready for inspection.

Notes and Examples

Seeds which are moist can be bruised and damaged during extraction. Seeds which are too dry are brittle and can easily crack. Cleaning by hand is gentler and will do less damage to the seeds.

With small samples it is more convenient and quicker to clean by hand because of the time needed to clean machinery between accessions. The flat bamboo baskets used in many countries are ideal for cleaning and winnowing.

Equipment:

- Small containers and trays
- Set of sieves
- Bamboo baskets
- Paper packets
- Threshing machine
- Seed blower
- Spatulas

Seeds from fleshy fruits:

1. Some species have fruits which irritate the skin, e.g. Solanaceae, and gloves should be worn when handling them.
2. Open the fruit using a knife but take care not to cut through the seeds.
3. Extract the seeds from each fruit and place in a tray.
4. Remove any juice or flesh adhering to the seeds using clean cold water and a small brush to remove tenacious flesh.
5. If mucilage surrounds the seeds and cannot be removed by washing in clean water, gently rub the wet seeds with coarse sand and then wash

off the sand and mucilage in a sieve. It is also possible to dry the seeds and then rub the dry mucilage off, but the seeds must be well separated, to prevent their sticking together during drying, on a surface to which they will also not stick.

6. Do not allow seeds to remain in the water too long because they will absorb moisture which can cause the start of germination and then cause a loss of viability when the seeds are dried.
7. Dry the seeds in a thin layer, preferably one seed thick, on cotton towelling or paper in the drying room, if available, or in a cool shady place.

Notes and Examples

Use your knowledge of the morphology of the fruit and the position of the seeds to ensure that seeds are not damaged by cutting. It is often extremely difficult to remove flesh from small seeds. It may be easier to dry the seeds and rub off the flesh when dry. Make sure that the paper or cotton towelling used is clean to prevent contamination and avoid affecting the subsequent germination.

- Equipment
- Plastic bowls
- Nail brush
- Absorbent paper
- Paper packets
- Spatulas

Examine the Cleaned Seeds for Insect and Fungal Damage

1. If the seeds are infested with insects or infected with fungi, isolate the sample and dry the infested seeds in sealed containers with silica gel to prevent the spread of the insect.
2. It is a general rule that drying seeds to a low moisture content will prevent further growth of fungi and development of further reproductive cycles of insects, but it will not kill them.
3. Storage at sub-zero temperatures will kill the adult insects but not fungi. Larva and eggs are more difficult to kill.
4. If seeds are to be stored above 0°C, place in a freezer for about 7 days to kill the insects before continuing with the normal packaging and storage procedures.

Notes and Examples

The use of low temperatures prevents fungal growth, but the spores remain viable during storage and can begin to grow again when the seeds are sown or germinated. To avoid this, fungicides can be used at planting but should not be used during storage because some chemicals can reduce seed longevity over long periods.

Effective treatment to kill insects without the use of insecticides which may damage seeds during storage is also possible. Seed drying prevents further development and sub-zero temperatures for about 7 days will kill the insects and larvae. Seeds must be dry or viability will be lost by treatment at these temperatures. Large seed or seeds packaged in bulk may take longer to treat to ensure that the centres of the seeds have been exposed to the low temperature.

Examine the Cleaned Seeds for Physical Damage and Empty Seeds

1. Spread the seeds on a flat well lit surface of contrasting colour. An illuminated table can be used if available.
2. Examine the seeds carefully for physical damage and empty seeds.
3. Discard any visually damaged, shrivelled or empty seeds.
4. Empty seeds may be difficult to detect unless the seeds are dissected or tested with special X-ray equipment to determine the percentage of empty seeds (Ch. 8, Ellis, Hong and Roberts, 1985).
5. Empty seeds are lighter than full ones and can be separated by use of the seed blower.

Notes and Examples

Take care that lightweight full seeds are not discarded during this process. If in doubt dissect some of lighter seeds taken from the debris to check whether they contain embryos.

Destroy the Waste Material

1. Destroy any waste material to prevent spread of the disease or insects to other material.

Notes and Examples

If possible burn the waste material to ensure complete destruction.

Seed Moisture Content

What is Seed Moisture Content?

The moisture content is the amount of water in the seed and is usually expressed as a percentage. It can he expressed on either a wet weight basis (where it is expressed as a percentage of the fresh weight of the seed) or on a dry weight basis (where it is expressed as a percentage of the dry weight of the seed). For genebank work it is usually expressed on a wet weight basis and all determinations and calculations should follow this rule.

The seed moisture content can either be accurately determined experimentally by scientific techniques, or it can be predicted approximately from the information available. The determination is destructive to the seeds used and since in many cases a prediction is sufficient, determination should only be carried out where essential.

Why is It Important to Determine Moisture Content?

A small change in seed moisture content has a large effect on the storage life of the seeds. Therefore it is important to know the moisture content in order to make a reasonably accurate prediction of the possible storage life of each accession.

When should Moisture Content be Determined and When should it be Predicted?

Accurate determinations should be kept to a minimum because for many activities a prediction is sufficient. In most cases this means one accurate determination after the drying period, so that the moisture content of the stored seeds is known and can be used to make better predictions of storage life.

How should Moisture Content be Determined and How should it be Predicted?

The methods outlined in the following pages can be used as a guide to prediction but experience in dealing with a crop in your own genebank will be necessary for reasonable predictions to be made.

The methods for determination are modified from those of the International Seed Testing Association (ISTA) as recommended by the IBPGR Advisory Committee on Seed Storage. Other methods are available and can also be used provided that the results give an accurate determination of moisture content and are calculated on a wet weight basis, to enable comparisons to be made between materials and between genebanks.

Prediction of Seed Moisture Content

The possibility of making reasonably accurate predictions of moisture content will improve as you gain experience with seeds of different species and use your knowledge of the origin of the seeds and the conditions in which they have been stored.

The moisture content of a seed will equilibrate with the relative humidity of the air surrounding it. This moisture content is called the equilibrium moisture content and because it is constant for any species at known temperature and relative humidity, it can be used as an approximation of the actual seed moisture content.

Moisture contents will differ from seed to seed within the same accession, but should not be sufficiently different to make more than about 1% difference in moisture contents between seeds of the same accession and different accessions of the same species. This is sufficiently close for a prediction. A list of a few common species and their equilibrium moisture contents is provided for guidance. Further information is available from Cromarty, Ellis and Roberts (1982). For species not on the list, it is possible to experimentally determine equilibrium moisture contents for these species at the temperatures and relative humidities at which you are working and prepare your own table to use for future predictions.

Determination of seed equilibrium moisture content

1. Do not waste seeds of germplasm for this, but use either excess seeds or those which are being discarded because they have lost viability.
2. Take two lots of seeds of the species of known weight and leave them in open containers in the environment of your genebank to equilibrate. This may be several days for small seeds and longer for large seeds. Seeds will have reached equilibrium with the moisture content of the air when there is no change in weight on different days.
3. Measure the relative humidity of the atmosphere where the seeds have been held for the past few days by using an aspirated hygrometre.
4. Measure the mean temperature.
5. Remove a sample of seeds and do an accurate determination of moisture content for each seed lot.
6. The mean of the two tests can be used as a guide because other seed lots of the same species should equilibrate at a similar moisture content under these conditions.
7. Repeat the determination the following day to make sure that the moisture content is stable and that the equilibrium has been reached,
8. The work can be repeated using a number of species which are of interest at a range of relative humidities that are common in your country at different times of the year.
9. Make a table of the equilibrium moisture contents at different temperatures and relative humidities and use this to refer to in the future.

Table: Equilibrium moisture contents of some common crop seeds at 25°C [1]

Species	***Percentage Relative Humidity***					
	15	30	45	60	75	90
Barley (Hordeum)	6.0	8.4	10.0	12.1	14.4	19.5

Beet (Beta)	5.8	7.6	9.4	11.2		
Buckwheat (Fagopyrum)	6.7	9.1	10.8	12.7	15.0	19.1
Cabbage (Brassica)	5.4	6.4	7.6	9.6		
Carrot (Daucus)	6.8	7.9	9.2	11.6		
Cucumber (Cucumis)	5.6	7.1	8.4	10.1		
Egg plant (Solanum)	6.3	8.0	9.8	11.9		
Flax (Linum)	4.4	5.6	6.3	7.9	10.0	15.2
Groundnut (Arachis)	2.6	4.2	5.6		9.8	13.0
Lettuce (Lactuca)	5.1	5.9	7.1	9.6		
Lima bean (Phaseolus)	7.7	9.2	11.0	13.8		
Maize (Zea)	6.6	8.4	10.2	12.7	14.4	18.8
Mustard (Brassica)	4.6	6.3	7.8	9.4		
Oat (Avena)	5.7	8.0	9.6	11.8	13.8	18.5
Okra (Abelmoschus)	8.3	10.0	11.2	13.1		
Onion (Allium)	8.0	9.5	11.2	13.4		
Radish (Raphanus)	5.1	6.8	8.3	10.2		
Rice (Oryza)	5.6	7.9	9.8	11.8	14.0	17.6
Rye (Secale)	7.0	8.7	10.5	12.2	14.8	20.6
Sorghum (Sorghum)	6.4	8.6	10.5	12.0	15.2	18.8
Soyabean (Glycine)	4.3	6.5	7.4	9.3	13.1	18.8
Tomato (Lycopersicon)	6.3	7.8	9.2	11.1		
Turnip (Brassica)	5.1	6.3	7.4	9.0		
Watermelon (Citrullus)	5.1	6.3	7.4	9.0		
Wheat (Triticum)	6.5	8.5	10.4	12.1	14.6	19.8
Winter squash (Cucurbita)	5.6	7.4	9.0	10.8		

Information Derived from: Roberts, E.H. (ed.) (1972) Seed Viability. Chapman and Hall, London.

How to Predict Seed Moisture Content

Measure Environmental Conditions

1. Measure the relative humidity of the atmosphere where the seeds have been held for the past few days by using an aspirated hygrometre.
2. Measure the mean temperature where the seeds have been held for the past few days.
3. If the seeds are from a different environment, they must be allowed to equilibrate with the current one before any predictions are made.

Notes and Examples

Wet and dry bulb thermometres can also be used to measure relative humidity. They do not give as accurate a result because the air around the bulbs is static. Allow sufficient time for the seeds to reach equilibrium. Large seeds take longer to equilibrate.

Equipment

Aspirated hygrometre

Maximum and minimum thermometre

Check the Equilibrium Moisture Content in the Table

1. Find the name of the crop in the left hand column of the tables of equilibrium moisture contents.
2. Read off the equilibrium seed moisture content at the relative humidity measured in Step 1 above.
3. Use this figure of approximate moisture content as your predicted value in making estimates of drying period or when an exact value of moisture content is not necessary.

Determination of Seed Moisture Content

Seed moisture content can be experimentally determined by various methods. The IBPGR Advisory Committee on Seed Storage have advised that, for use in genebanks, the ISTA methods are accurate and acceptable. However, considering that the cost of germplasm is high and moisture content determination is destructive, it has been recommended that the ISTA methods should be modified by reducing the weight of seeds required for each replicate whilst increasing the accuracy of the weighings.

The ISTA has only recommended standard methods for common crop seeds. There are therefore no standard methods for many of the species stored in genebanks. The ISTA methods can be applied to seeds of these other species where appropriate as explained in the following pages.

Species whose moisture content should be determined by ISTA methods.

Modified low constant temperature oven method:

- Brassicas
- Castor (Ricinus)
- Pepper (Capsicum)
- Cotton (Gossypium)
- Eggplant (Solanum)
- Falseflax (Camelina)

- Flax (Linum)
- Groundnut (Arachis)
- Onion (Allium)
- Radish (Raphanus)
- Sesame (Sesamum)
- Soyabean (Glycine)
- All tree species

Modified high constant temperature oven method:

- Alfalfa (Medicago)
- Asparagus (Asparagus)
- Barley (Hordeum)
- Bean (Phaseolus)
- Beet (Beta)
- Bentgrass (Agrostis)
- Bermuda grass (Cynodon)
- Black salsify (Scorzonera)
- Bluegrass (Poa)
- Brome (Bromus)
- Buckwheat (Fagopyrum)
- Canarygrass (Phalaris)
- Caraway (Carum)
- Carrot (Daucus)
- Chervil (Anthriscus)
- Chickory (Cichorium)
- Chickpea (Cicer)
- Clover (Trifolium)
- Cocksfoot (Dactylis)
- Cress (Lepidium)
- Crested dogtail (Cynosurus)
- Cucumber (Cucumis)
- Cumin (Cuminum)
- Dallisgrass (Paspalum)
- Fescue (Festuca)

- Foxtail (Alopecurus)
- Lettuce (Lactuca)
- Lupin (Lupinus)
- Maize (Zea)
- Millet (Panicum)
- Oat (Avena)
- Parsley (Petroselinum)
- Pea (Pisum)
- Rhodes grass (Chloris)
- Rice (Oryza)
- Rye (Secale)
- Ryegrass (Lolium)
- Sainfoin (Onobrychis)
- Serradella (Ornithopus)
- Sorghum (Sorghum)
- Squash (Cucurbita)
- Sweetclover (Melilotus)
- Tall oatgrass (Arrhenatherum)
- Timothy grass (Phleum)
- Tomato (Lycopersicon)
- Trefoil (Lotus)
- Tufted hairgrass (Deschampsia)
- Velvetgrass (Holcus)
- Vetch (Vicia)
- Watermelon (Citrullus)
- Wheat (Triticum)

How to Determine Seed Moisture Content

Notes and Examples

Great care is required especially with small samples when doing the determination to ensure that the results are valid. In order to reduce experimental error, light-weight dishes can be made from strong aluminium foil, but these are fragile to handle and covers are not easy to make.

Always work with care and finish one sample at a time. Do not leave the dishes open in the laboratory between weighings because the seeds will either lose or absorb water from the air and small changes in weights can result in large differences in the calculations when the amount of seed used is small.

Determine If Pre-Drying is Required

1. Use your experience to predict moisture content as explained in the previous Section.
2. If the seeds are very moist, pre-drying may be necessary. ISTA requires pre-drying for seeds of the species which require grinding listed in Table 2 of more than 17% seed moisture content, for seeds of soyabean of more than 10% moisture content, seeds of rice of more than 13% moisture content and seeds of other species of more than 30% moisture content.

Determine If Grinding is Necessary

1. Check the ISTA list of species to see if the seeds require grinding.
2. If the seeds are not on the list but are larger than wheat seeds, grinding is required.

Notes and Examples

Grinding should be done carefully to prevent errors in moisture content determination through water loss or absorption during the process. Hard seeds can be difficult to grind and a pestle and mortar may be useful. Very moist seeds will also be difficult to grind and a sharp knife and chopping board can be used to cut them into small pieces of the recommended sizes.

Notes and Examples

High temperatures cannot be used to determine the moisture content of oily seeds because the oil will also vaporize and give a false result of water plus oil content. Temperatures of just over 100 °C allow evaporation of water and minimal vaporization of oils. If you are in doubt as to which method to use, choose the low temperature method for safety.

Modified ISTA Low Constant Temperature Oven Method

1. If pre-drying is not required, weigh out approximately the total weight of seeds to be used, as determined in Step 1.
2. If pre-drying is required, weigh out accurately two sub-samples of between 1 - 1.5 g of seeds. Pre-dry these samples in a warm dry place overnight. Weigh again, writing all the weights in the note book.
3. If grinding is necessary, grind the seeds of each sample, directly before weighing, into pieces not larger than 4 mm for legumes and tree seeds and 0.5 - 1 mm for cereals.

4. Keep the ground seeds in a covered dish and go directly to the weighings.
5. Weigh one clean numbered dish and cover accurately to 4 decimal places using an analytical balance. Write the weight in the notebook.
6. Add approximately 0.5 g of the prepared seeds distributed evenly over the base of the dish, replace the cover and accurately weigh the dish and cover and write this weight in the notebook.
7. Place the dish in a safe place and continue to do the second replicate and additional samples in the same way.
8. When all samples have been weighed into numbered dishes, place each dish on top of its numbered lid in the oven at 101 - 105 °C.
9. Wait for the oven to reach this temperature and heat the samples for 15 - 17 hours.
10. Remove the dishes from the oven, replace their covers and place in a desiccator to cool for 30 to 45 minutes at room temperature.
11. Remove the dishes one by one from the desiccator and immediately weigh each dish and cover and note the weight in the notebook. Do not leave the desiccator open during the weighings.

Notes and Examples

Use a standard table form to record the results and simplify your work and calculations. After heating, make sure that the dishes are put directly in the desiccator so that the dry seeds do not absorb more moisture. Do not try to close a desiccator with hot dishes inside until the air has expanded, or the expansion of air can break the seal and cause the lid to slide off. After cooling there will be a slight vacuum in the desiccator and the lid may be difficult to remove. Grease the lid of the desiccator to make it easier to remove. The use of a desiccator with a tap in the lid can prevent these problems.

Modified ISTA High Constant Temperature Oven Method

1. If pre-drying is not required, weigh out approximately the total weight of seeds to be used, as determined in Step 1.
2. If pre-drying is required, weigh out accurately two sub-samples of between 1 - 1.5 g of seeds. Pre-dry these samples in a warm dry place overnight. Weigh again, writing all the weights in the note book.
3. If grinding is necessary, grind the seeds of each sample, directly before weighing, into pieces of not larger than 4 mm for legumes and tree seeds and 0.5 - 1 mm for cereals.
4. Keep the ground seeds in a covered dish and go directly to the weighings.
5. Weigh one clean numbered dish and cover accurately to 4 decimal places using an analytical balance. Write the weight in the notebook.

6. Add approximately 0.5 g of the prepared seeds distributed evenly over the base of the dish, replace the cover and accurately weigh the dish and cover and write this weight in the notebook.
7. Place the dish in a safe place and continue to do the second replicate and additional samples in the same way.
8. When all samples have been weighed into numbered dishes, place each dish on top of its numbered lid in the oven at 130 - 133 °C.
9. Wait for the oven to reach this temperature and heat the samples for 2 hours for cereals, 4 hours for maize and 1 hour for all other species on the ISTA list.
10. Remove the dishes from the oven, replace their covers and place in a desiccator to cool for 30 to 45 minutes at room temperature.
11. Remove the dishes one by one from the desiccator and immediately weigh each dish and cover and write the weight in the notebook. Do not leave the desiccator open during the weighings.

Seed Drying

What is Meant by Seed Drying?

This is the reduction of seed moisture content to the recommended levels for seed storage, using techniques which will not be detrimental to seed viability.

Why are Seeds Dried?

Seeds which are dry will retain their viability for longer periods of storage in genebanks. It is recommended that, in general, seeds should be dried to between 3-7% moisture content for long-term storage, except in certain cases where it has been shown that low moisture content causes problems e.g. soyabean should be dried to about 8% moisture content. Drying has an independent effect from temperature on viability during storage and adequate drying could prolong viability for reasonably long periods without cold storage.

When should Seeds be Dried?

The drying process should begin as soon as possible after the receipt of the seeds to avoid unnecessary deterioration. It can take up to several weeks to lower the moisture content to the low levels required for good storage.

Prediction of Drying Period

If there is no previous experience of drying seeds of a particular species it may be necessary for you to do some experimental work to predict the approximate drying period. Seeds dry at an exponential rate until the equilibrium moisture content is reached. The rate of drying of different seed lots of the same species is fairly constant for seeds dried under the same environmental conditions. The length of the drying period can be predicted in one of two ways:

Prediction of the Correct Drying Period by Weight Loss

1. Predict the current percentage moisture content and the percentage moisture content required for storage.
2. Weigh the seed sample.
3. Use these three values to calculate the weight of seeds at the required moisture content by using the following formula:

$$Final\,seed\,weight \quad Inital\,seed\,weight \; \frac{(100 \quad Initial\%moisture\,content)}{(100 \quad Final\%mosture\,content)}$$

4. Weigh the seed sample at regular intervals during the drying period until the weight of the seeds has reached this calculated value.

Notes and Examples

The prediction of the drying period will be more accurate if the initial moisture content is determined. However, this is a waste of seeds when moisture content can be adequately predicted.

1000 g of seeds with 12% moisture content were dried to 5% moisture content. What would be the weight of these seeds after drying?

Substitute the values in the equation on the facing page:

$$Final\,seed\,weight \quad 1000 \; \frac{(100 \quad 12)}{100 \quad 5} g \quad 926.3g$$

Therefore when the initial 1000 g of seeds have been dried to 926.3 g, their moisture content will have decreased from 12% to 5%.

Prediction of drying period from mean drying curves

1. Do not waste valuable seeds for this, but use either excess seeds or those which are being discarded because they have lost viability.
2. Take at least two lots of seeds of the species and place to dry, using the method that you will use in practice.
3. Remove a sample of seeds and do an accurate determination of moisture content for each seed lot as described in Section III.
4. The mean of the two tests can be used as a guide because other seed lots of the same species should dry at a similar rate.
5. Repeat the determination daily and plot a graph of the drying curve (mean percentage moisture content against time) for that species under these drying conditions. Your graph will be similar to the one on page 44.

6. The work can be repeated with seeds of all species which are of interest and their drying curves plotted for different conditions.

Notes and Examples

If the seeds are very small and are dried quickly, the moisture content can be determined more frequently during the drying period to plot a more accurate graph.

How to use the drying curves

1. Use the graph that you have prepared for seeds of a particular species being dried under those conditions.
2. Predict the current percentage moisture content of the accession by using the table of equilibrium moisture contents. Select the final percentage moisture content that is required for storage of this species.
3. Find the value for each of the percentage moisture contents on the vertical 'Y' axis, follow this across to the curve and read off the day on the horizontal 'X' axis for each of the moisture contents. The difference between these two values is the drying period in days.

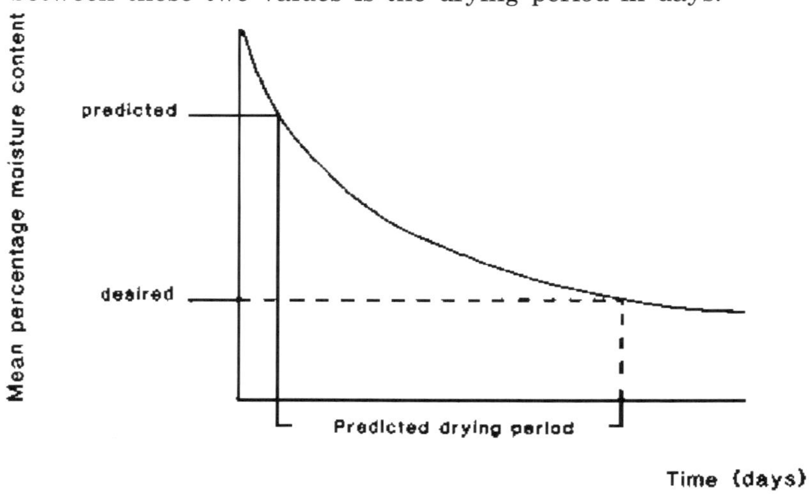

Figure: *Typical drying curve*

Notes and Examples

Example of how to predict the drying period by using a typical drying curve for small seeds of crops such as onion or cabbage, dried quickly under low relative humidity conditions: Predict the moisture content by using the table of equilibrium moisture contents for your genebank which you have prepared. As an example take this as 12% and the desired moisture content after drying as 5%. Using the graph, the intersects from the curve to the time axis show the values 10 and 2.25 days.

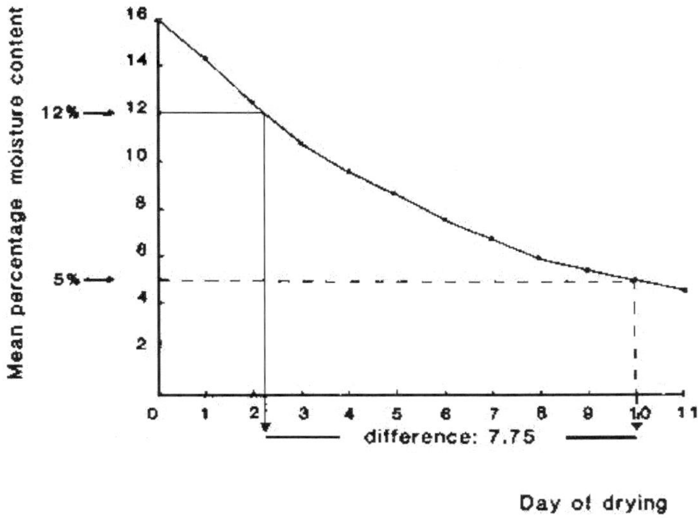

Figure: *Typical drying curve of small seeds*

Therefore, the time to dry the seeds to 5% moisture content under the same conditions will be the difference between 10 and 2.25 days. This can be approximated to the nearest day, in this case 8 days.

In practice it is unlikely that using the graph will give you a whole number of days, therefore you should approximate to the nearest day.

Graphs of Mean Drying Curves in Your Genebank

1. Species:
   - Temperature
   - Relative humidity

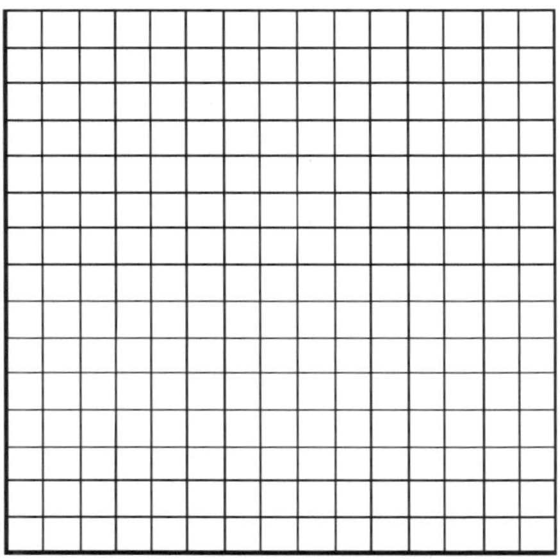

Plot graphs in the space below to use for future reference:

2. Species:
   - Temperature
   - Relative humidity

3. Species:
   - Temperature
   - Relative humidity

4 Species:

- Temperature
- Relative humidity

5. Species:

- Temperature
- Relative humidity

6. Species:
    - Temperature
    - Relative humidity

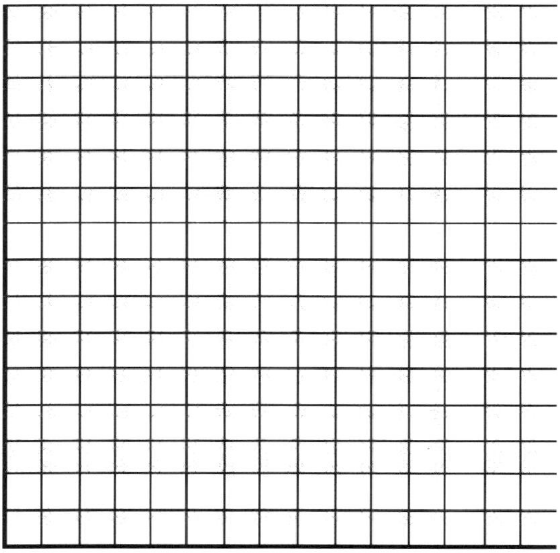

7. Species:
    - Temperature
    - Relative humidity

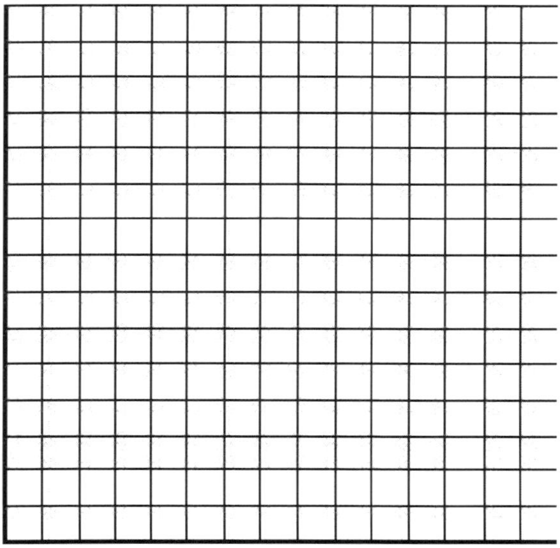

8. Species:

- Temperature
- Relative humidity

8

Seed Drying Procedures

Predict Moisture Content and Drying Period

1. Use the experience and techniques already described in Section III to predict the moisture content of the accessions.
2. Predict the length of drying period required from your previous experience of drying seeds of the same species with similar moisture contents, as already described. Information on drying periods can also be found in Cromarty, Ellis and Roberts (1982).

Notes and Examples

Predict the moisture content of each accession and the number of days to reach the required moisture content. Remember that seeds will dry more slowly as the moisture content decreases towards the equilibrium moisture content. At low moisture contents a difference of one or two days of drying will not have a large effect on seed moisture content.

Prepare The Seeds For Drying

1. Place each accession in labelled paper envelopes or cloth bags.
2. Do not put too many seeds in the same bag. One accession can be split into several labelled bags to help rapid drying.

Dry The Seeds

Several methods are available for drying seeds. The safest methods rely on leaving the seeds in an environment of low relative humidity and allowing the seed moisture content to reach equilibrium with this at relatively low temperatures. Seeds will equilibrate with the relative humidity at different rates, depending on species, seed size and conditions. Seeds dry quicker when first placed in the room and more slowly as the low moisture contents suitable for storage are approached. Two common and safe methods are described here:

Notes and Examples

Choose the best drying method for your own genebank. In tropical countries with high relative humidities, it will be more difficult to hold a drying room at very low relative humidity. A combination of methods is possible. Silica gel can be used after the first drying period if the relative humidity is too high to allow the seeds to equilibrate to low moisture contents.

Do not use sun drying because it is believed to affect long term seed viability of some species. Heated air drying should also be avoided for the same reason.

Aim for 10 - 15% relative humidity and 15°C as the optimum drying conditions.

Use of a dehumidifier dryer:

1. Use a dehumidifier dryer to dry a small volume of air in a drying room or other limited space. The humidity of this room will depend on its size, the ambient humidity of the area and the efficiency of the dehumidifier.
2. Place the seeds which have been packaged in labelled porous containers in the drying area.
3. Do not stack the bags too closely and use open racks in a drying room or cabinet with a fan to allow air to circulate.
4. Measure the relative humidity and maximum and minimum temperatures of the room daily.
5. Leave the seeds in the drying area until the moisture content is predicted to be in the range required for storage.

Use of Silica Gel

1. Use deep blue silica gel in an enclosed space, such as a desiccator, can or glass jar with an air-tight seal.
2. Place either the silica gel or the seeds in a porous bag.
3. Use a weight of silica gel equal to the weight of seeds for rapid drying.
4. Place the required weight of silica gel in a desiccator, jar or can with the correct weight of seeds. When using cans or jars make sure that the silica gel is in close proximity to, but not touching, the seeds.
5. Place the containers in a room held at approximately 15 °C whilst the seeds are drying.

6. Change the silica gel daily or when the colour changes from deep blue to pale blue or pink.
7. Heat the pale blue or pink silica gel in an oven above 100 °C until it turns deep blue again, when it is ready for re-use. Store in an air-tight container.
8. Leave the seeds with fresh changes of silica gel in the container until the moisture content is predicted to be in the range required for storage.

Notes and Examples

Desiccators are useful but expensive. Cans and jars with air-tight seals are cheaper and adequate and also allow separation of seeds of different accessions. Silica gel is cheap and easy to use. It can be made into permanent packages of fixed weight in fine cloth bags. These can be heated in the oven without damage and are quicker and easier to handle than loose silica gel. The silica gel is used to reduce the relative humidity of the atmosphere surrounding the seeds. The seeds then dry by equilibration with their surroundings.

Accurately Determine the Moisture Content

1. When you consider, from experience, that the seeds are dry enough, remove a sub-sample from each accession and carry out a moisture content determination as explained in Section III.
2. When the moisture content is between 3-7 % for most species, go to packaging (Section VI). Low moisture contents are detrimental to the viability of the seeds of some crops. These seeds should be treated with care and not dried to low moisture levels, e.g. soyabean should not be dried below 8% moisture content.
3. If the moisture content is not low enough, continue to dry for a further period as described above. When the moisture content is predicted to be in the correct range for seed storage, carry out an experimental determination as described in Section III.

Notes and Examples

At this point it is necessary to do an experimental determination of moisture content to ensure that predictions of storage life of this accession are as accurate as possible. Make sure that seeds which have been dried are kept in the drying area or moisture-proof containers whilst tests are carried out to prevent absorption of moisture from the surroundings.

Seed Packaging

What is Meant by Seed Packaging?

This is the placing of a counted or weighed sample of seeds of an accession into a container which is then hermetically sealed ready for storage.

Why are Seeds Packaged?

Seeds are packaged to prevent absorption of water from the atmosphere after drying, to keep each accession separate and prevent contamination of the seeds from insects and diseases.

When should Seeds be Packaged?

The best time to package seeds is directly after the moisture content has been determined and found to be within the required limits for safe storage. Seeds will always show an equilibrium between their moisture content and the relative humidity of the environment and therefore, if possible, seeds should be packaged into containers and hermetically sealed in the drying room or without delay on being removed from it.

How should Seeds be Packaged?

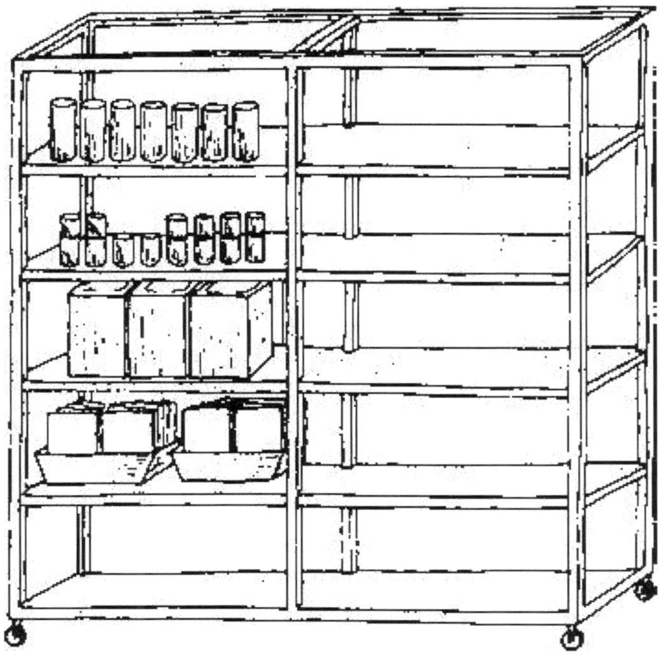

Different containers and special equipment for sealing are available for the storage of seeds. Storage containers for base collections should be hermetically sealed and moisture-proof. Cans, bottles, and laminated aluminium foil containers are all acceptable for both base and active collections. The techniques used will vary with the type of container and the equipment that your genebank is using. The general steps outlined in this section could be followed. Moveable racks make the best use of available space and are ideal to store containers in walk-in stores. Small containers or aluminium foil packets can be filed in boxes for ease of locating individual accessions. Coding systems by number or colour are also helpful in exactly locating accessions. Chest freezers are adequate for the maintenance of small collections. Filing systems have been developed to hold the containers in freezers and allow easy location of material. Some examples of filing systems are shown here.

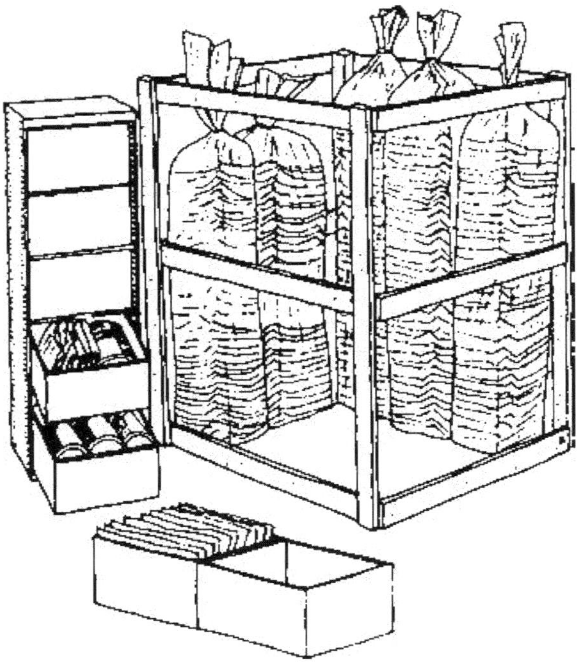

Prepare for Packaging

1. Work in the drying room or, if not possible, expose the seeds to the ambient relative humidity for the shortest possible time.
2. Write on the outside of each container or on an adhesive waterproof label. Also prepare a label for inclusion with the seeds. Record the accession number, date of storage, genus and species if required. Use permanent markers for this.

Notes and Examples

Adhesive labels can be used for the outside of containers, but they must be waterproof and remain adhesive for long periods at low temperatures.

Equipment

Labels

Permanent markers

Laminated aluminium foil containers, cans or bottles

Machines for sealing

Coarse balance

Scoop/spoons

Package the Seeds

1. Weigh out or count samples of seeds to fit the containers used in your genebank.
2. Fill the labelled containers with the seeds. Add the label prepared for the inside.
3. Seal immediately, so that the moisture content of the seeds does not increase due to equilibration with ambient relative humidity.
4. Note the weight or number of seeds in each container.

Notes and Examples

It is important that the containers used should be moisture-proof and sealed. The exact sealing technique will depend on the type of containers and sealing methods that are available. Laminated aluminium foil bags are easy to package, can be cut to size to save space and can be sealed again after use. However, they are difficult to stack and must be made of good quality material and have good seals or leakage may occur.

Cans are rigid and easy to stack and usually will not break open if dropped. However, some types of cans are not resealable and therefore are expensive to use. A standard size can with only a few seeds inside wastes space in the store. Glass bottles can usually be sealed again and the amount of seed left is visible. However, a standard size will waste space and bottles are easily broken.

Seeds can be packaged in bulk into large containers and/or sub-samples can be packaged separately into smaller containers. Although the initial packaging period is longer, sub-samples can then be removed quickly without having to remove the bulk of the accession from the store.

Check the Quality of the Containers

1. After sealing, make a visual examination of each container to make sure that there is no obvious damage and that the seals do not leak.
2. Any containers that are below standard should be replaced immediately.
3. At regular intervals the containers should be checked to see that they remain in good condition. It is suggested that this check should be carried out routinely once a year and that individual containers should also be checked whenever they are removed from the genebank.

4. If containers are found to have been leaking and the relative humidity of the store was not controlled, determine the seed moisture content by using one of the methods described in Section III.
5. If the moisture content has risen, dry the seeds back to the required moisture content level as described in Section IV.
6. Enter the value of the new moisture content into the data files. Make a note that the seeds in that container have been held at increased moisture content for a limited period and dried again.
7. Check the inventory data file for the descriptor 'date of packaging' and make a list of any other accessions which were packaged in similar containers on the same day or one day before or after.
8. Check the containers on this list for leaks and poor condition and replace any that are faulty using the methods described above.
9. Remember that any containers removed from the cold store should be allowed to warm to room temperature before opening. This may take several hours especially with large volumes of seeds.

Notes and Examples

If any defective containers are found, it may indicate that containers made or sealed at the same time are also faulty or it may just be one faulty container or seal. Checking those packaged at the same time will show if the problem is widespread.

Seed Storage

What is Meant by Seed Storage?

In a genebank, seed storage is the preservation of seeds under controlled environmental conditions which will prolong the viability of the seeds for long periods. Two types of seed stores are used for collections of genetic resources, those holding base collections and those holding active collections. The temperature, relative humidities, seed moisture contents, containers and distribution arrangements vary between these stores.

Why are Seeds Stored?

Seeds must be stored in a way which maintains their viability for long periods. Seeds left at ambient temperatures and relative humidities will lose their viability quickly whilst seeds stored in conditions of low moisture content and temperature will retain their viability for longer periods. Accessions held in a genebank are valuable and represent plants which are no longer available or which are endangered in their natural environment. These seeds must be conserved in the genebank for use in plant breeding in the future.

When should Seeds be Placed into Store?

As soon as the seed has matured on the plant the slow process of deterioration begins. Therefore the sooner that seeds are placed into store the better. In practical terms this means that seeds collected in the field should be quickly returned to genebanks, processed and placed into store as soon as the cleaning, drying and packaging processes are complete.

How many Seeds of Each Accession should be Stored?

The IBPGR Advisory Committee on Seed Storage has recommended that for materials showing little morphological variation (genetically homogeneous) 3000 seeds are acceptable, but 4000 seeds are preferred, to represent each accession. For materials showing a large amount of morphological variation (heterogeneous) an accession should consist of at least 4000 seeds, but 12000 seeds are preferred. These sample sizes are recommended as number of seed but many genebanks would prefer to work with weight. An inter-conversion of numbers and weights is possible from the thousand seed weight of any accession. A list of approximate thousand seed weights of some common species is given in Appendix 2 of Cromarty, Ellis and Roberts (1982).

How should Seeds be Stored?

The conditions which prolong viability during storage have been well defined for seeds which are tolerant of desiccation. Storage conditions have been recommended by the IBPGR Advisory Committee on Seed Storage. For base collections, seeds of between 3-7 % moisture content should be stored in sealed containers. Sub-zero temperatures are acceptable, but -18 °C or less is preferred. For active collections sealed storage of seeds dried to 7% moisture content or less is recommended at temperatures of less than 15 °C. Unsealed storage is not encouraged. In particular, it is not recommended in tropical areas.

Check the Number of Seeds in the Accession

1. Weigh the seeds of each accession.
2. Convert the weight of seeds to seed number, using the thousand seed weight of each accession for an accurate conversion. An approximate conversion can be done on a species basis by using the approximate weights given in Appendix 2 of Cromarty, Ellis and Roberts (1982).
3. For accessions containing mixtures of genotypes, the sample size should be at least 4000 seeds. For genetically uniform accessions, the sample size should be at least 3000 seeds.
4. If the sample contains more than the required number of seeds, proceed to storage.

5. If the sample contains less than the required number of seeds, either proceed directly to regeneration or store the seeds temporarily in the genebank and regenerate at the earliest opportunity.

Notes and Examples

An approximate inter-conversion of seed number and weight can be done easily using the thousand seed weight.

Example for Sorghum (1000 seed weight = 17 g)

$$100g\ contains: \frac{1000}{17}\quad 100\quad 5750 seeds.$$

These are minimum sample sizes for the start of storage and if both space and seeds are available, more seeds should be held. Regenerate as soon as possible. Seeds processed and stored under good conditions will not loose viability before regeneration.

Determine Where the Seeds should be Located

1. Check the inventory data file of the genebank to find the next available space where a container can be located.
2. When seeds from the same regeneration cycle of the same accession are stored in several containers, keep all the containers of the accession together.
3. Make a list of where each accession will be placed.

Notes and Examples

A code can be used to locate an accession within the store. Each unit can be identified by a number or letter in the store. The code can indicate the number or the letter of the freezer, store, rack, basket or drawer, etc.

Colour codes can also be a quick and easy way to locate accessions. A colour can be used for each rack, shelf or species. This both speeds up the work in the coldroom and makes it easy to spot errors. Owing to the very cold temperatures, the faster that one can locate accessions in the cold room the better.

Monitoring Accessions

What is Meant by Monitoring?

As previously explained, the viability of the seeds will decrease during storage over long periods. The number of seeds of each accession held will also decrease as seeds are removed for testing and in the case of active collections, when seeds are distributed. Both viability and seed number must be monitored during storage.

Why must Accessions be Monitored?

Seed viability and number of seeds of each accession in the store must always be known with reasonable accuracy. Only then can accurate decisions on regeneration be made.

When should Seeds be Monitored?

The number or weight of seeds in store should be monitored every time seeds are removed from the genebank. Seeds should not be removed if only a few seeds remain. The viability should be monitored at regular intervals. This interval will be determined by the curator depending on the species, viability of the seeds at the start of storage, moisture content and storage conditions. The IBPGR Advisory Committee on Seed Storage recommends that seeds stored under the preferred standards in base collections should be monitored at least every ten years and that those with either poor storage life or poor initial viability in base collections, together with all seeds stored in active collections, should be monitored every five years.

How should Accessions be Monitored?

The number or weight of seeds can be monitored automatically by checking in the genebank management data file and need not be described further here. The viability can be monitored by doing a fixed sample size germination test as already described in Section V or using a sequential method as explained here. Both tests use the same germination conditions and differ only in the number of seeds used for the test and the interpretation of the results. On average the sequential germination test uses fewer seeds and the IBPGR Advisory Committee on Seed Storage has recommended it for monitoring collections in genebanks.

Determine Which Accessions Require Monitoring

1. The fact that accessions will be entered into the genebank in an irregular pattern, with many being entered at the same time after harvest or a collecting mission and few accessions being entered at other times, means that monitoring of accessions will also be in an irregular pattern.
2. Every week check the genebank management data file to see which accessions require monitoring during that week.
3. Prepare a list of all accessions by number, which require monitoring during the week.
4. If there are many accessions which require monitoring at the same time, make subgroups of about 20 accessions of similar species which can be handled with the same technique and do these on different days to spread the work.

5. If there are few accessions which require monitoring at the same time, it may be more efficient for them to be grouped and the monitoring done once a month.

Notes and Examples

Do not handle too many accessions at the same time. Availability of staff and equipment determine the number of accessions handled at one time. Samples needing monitoring within one or two weeks can be grouped together for convenience.

Prepare the Seeds for Testing

1. Check each accession number in the management data file to make sure that sufficient seeds are available for the test. If few seeds are available, do not test now because the material will have to be regenerated soon.
2. Make a list of each accession number, its location in the seed store and any data relating to the methods and conditions used for the germination test.
3. Remove the seed containers from the store and leave at room temperature, until the container has warmed up, before opening.
4. Count the number of seeds needed for the test from each accession. Work in a dehumidified room, if possible, and close the containers as soon as the seeds are removed to prevent the seeds absorbing moisture from the air.

Notes and Examples

Large containers may take several hours to reach room temperature, but it is important that cans should not be opened before room temperature is reached or condensation may form directly on the seeds and cause changes in moisture content.

Do a Germination Test

Two methods are recommended by the IBPGR Advisory Committee on Seed Storage:

Fixed Sample Size Germination Test

1. Do a fixed sample size germination test using a minimum of 200 seeds sown as two replicates of 100 seeds and follow the techniques already described in Section V.
2. Calculate the percentage germination from the results of the test.
3. If it is above 90%, accept the test as valid and use this value as the true viability.

4. If the result is 90% or below, repeat the test using a further 200 seeds following the same procedures.
5. Calculate the mean percentage viability from the results of the two tests and use this as the overall test result.
6. Compare the final result of this test with the results of the previous test and decide whether regeneration is necessary, according to the following criteria:

If the percentage germination has decreased to 85 or less, regenerate the accession. If the percentage germination is above 85, the curator will plan the next monitoring test date depending on the germination percentage.

Notes and Examples

The fixed sample size germination test is done using the standard techniques for the species in your genebank, as already explained in Section V.

Equipment

Filter paper, paper towelling or rice straw paper

Plastic bags

Petri dishes or other containers

Water

Forceps

Incubators

Labels

Permanent markers

Sand

Pots or trays

Tools for planting

Sequential germination test

1. A sequential germination test uses the same conditions and techniques as already described for the fixed sample size germination test, but uses less seeds for each replicate. The replication is done in a sequential way when necessary and thus not all replicates are tested at the same time.
2. The number of seeds required for each replicate can vary; the more seeds and replicates, the more accurate the test. It is recommended that a sample should consist of 40 seeds per replicate as a minimum.
3. Do a germination test using the conditions and methods as already described in Section V using one replicate of 40 seeds.

4. Count the number of seeds germinated.
5. Compare the results of the test with the number germinated in Table 3, which shows the germination test plan for the 85% regeneration standard using a sample size of 40 seeds per replicate.
6. Look at the line with the value of 40 in the first column (number of seeds tested).
   (i) If the number of seeds germinated is 29 or less, the accession will require regeneration as explained in Section X.
   (ii) If the number of seeds germinated is more than 29, then the test will have to be repeated with another sample of 40 seeds exactly as above.
7. It is important that each sample should be treated the same, so that the different samples can be used as replicates.
8. Count the number of seeds germinated in the second test and add this to the result of the number of seeds germinated from the first test. Add the total number of seeds used for both the samples.
9. Compare the results of the test with the number germinated in Table 3 as before. Look at the line with the value equal to the total number of seeds used for all tests (80) in the first column (number of seeds tested).
   (i) If the number of seeds germinated is 64 or less, then the accession will require regeneration.
   (ii) If the number germinated is between 65 and 75, the accession will require testing again with another sample of 40 seeds.
   (iii) If the number germinated is above 75, then the viability of the accession is still above the regeneration standard and the accession does not require regeneration.
10. Continue in this way until either a definite answer as to whether to regenerate or to continue storing the accession is reached or the test has been repeated ten times. The results should then be expressed as percentage germination and a decision as to whether to regenerate taken as already explained in the fixed sample size germination test monitoring method.

Notes and Examples

The sequential test has an advantage in that it uses less seeds but is of a similar accuracy to the fixed sample size germination test. The actual germination test conditions and procedures are the same for both types of germination test, the only difference being in the number of seeds used. The cost of storing germplasm is high; therefore, seeds should not be wasted. The cost of regeneration and the possibility of selection and loss of accessions during regeneration are high.

Therefore, test results should be accurate and indicate when regeneration is needed. A balance has to be found between using enough seeds to be accurate and using too many wastefully. This balance is found in the sequential germination test.

If seeds are thought to be of high viability or questionable, it may be simpler to set up the first two sequential replicates together (2 x 40) because as can be seen in Table 3, a second test will probably be necessary. It is easier and quicker to set up two tests together than to repeat the whole process after the results of the first test are known. There is then no delay in getting the results of the second test. When you know prior to testing that the seeds are probably in poor condition, one test is sufficient to show that regeneration must be done.

It is obvious that the test is open-ended and if the answer always fell within the range where the accession required testing again using another sample, a large amount of seeds would have to be used to determine whether regeneration is necessary.

Therefore it is suggested that in circumstances where no definite decision can be made regarding regeneration, the test should be terminated when ten samples have been tested. This uses 400 seeds, which is the same as in the fixed sample size germination test for accessions of less than 90% viability. Results can then be converted to percentage germination and a decision on the need for regeneration can be made.

- Equipment
- Filter paper, paper towelling or rice straw paper
- Plastic bags
- Petri dishes or other containers
- Water
- Forceps
- Incubators
- Labels
- Permanent markers
- Sand
- Pots or trays
- Tools for planting

Table 3. Modified sequential germination test plan for 85 percent regeneration standard for groups of 40 seeds

Number of seeds tested	***Regenerate if number germinated is less than or equals:***	***Repeat test if number germinated is between:***	***Store if number germinated is more than or equals:***
40	29	30 - 40	-
80	64	65 - 75	76
120	100	101 - 110	111
160	135	136 - 145	146
200	170	171 - 180	181
240	205	206 - 215	216
280	240	241 - 250	251
320	275	276 - 285	286
360	310	311 - 320	321
400[1]	340		341

When 400 seeds have been tested, the test can be terminated because enough tests have been done for a practical decision to be made.

Distribution of Germplasm from Active Collections

What is Distribution of Germplasm?

This is the supply of seeds of the different accessions in response to requests from plant breeders and research workers. Seeds are distributed only from active collections and not from base collections.

Why is Germplasm Distributed?

The purpose of keeping the germplasm safe in the genebank is to use it in plant breeding and related activities. It may be distributed for breeding, research, characterization and evaluation or for regeneration of fresh seeds of each accession.

When is Germplasm Distributed?

Germplasm is distributed freely from the genebank on request. However, considering the high cost of conserving it, the curator must feel that the request is justified.

Decide Whether the Accession should be Distributed

1. Check in the data file to see if seeds are available to distribute in response to the request. When the request has been filled there should remain in store a minimum of three times the number of seeds required for one regeneration cycle.
2. When there are not enough seeds, do not distribute. Inform the curator to programme regeneration and to write to the person who has requested the seeds to inform them that the accession cannot be supplied until after regeneration.

Notes and Examples

Seeds should not be distributed from base collections. Care must be taken not to distribute all available seeds of one accession. As many accessions as possible should be kept available for distribution, although it must be accepted that there will be a temporary delay in distribution until seeds can be regenerated.

Prepare the Sample for Distribution

1. Remove the container from the genebank.
2. Allow the container to warm to room temperature before opening (this may be several hours for large containers).
3. Write the accession number and genus and species of the seeds on two labels, one for the outside and one for inside the seed packet.
4. Work in a dehumidified room, if possible, and close the containers as soon as the seeds are removed to prevent the seeds absorbing moisture from the air.
5. Open the container and use a random sampling technique to weigh or count the required amount or number of seeds quickly. It is suggested that about 50 - 100 seeds should be distributed to fill each request.
6. Place the seeds with one label in an aluminium foil packet, seal and attach the other label to the outside.
7. Record the number or weight of seeds removed from each accession.

Notes and Examples

Always use random sampling techniques to remove seeds from the container so that the resulting sample will provide a good representation of the original population.

Prepare the Information Lists to Accompany the Seed

1. Prepare a list of all materials sent and their passport, characterization and evaluation data and any relevant genebank management data relating to the germination or dormancy behaviour of the seeds of each accession.
2. For overseas requests, obtain a phytosanitary certificate to accompany the seeds from the relevant authority. If certification of the seeds is not possible, inform the recipient and request him to advise on whether the seeds can be accepted into his country without the certificate.

Notes and Examples

Send as much information as possible with the seeds. Make sure that the phytosanitory regulations in both your country and the recipient country are met; otherwise seeds may be delayed or confiscated. Fungicides and insecticides can be used in order to comply with phytosanitory restrictions during distribution because it is expected that the seeds will be sown on arrival.

Despatch the Seeds

1. Despatch the seeds by air mail or air freight for overseas and by the most direct route for destinations in your country. This will avoid delay and damage to the seeds during transit.
2. Send one copy of the list and phytosanitary certificate with the seeds. Also send copies of the documents separately to the recipient.

Enter the Data into the Data Files

1. Record the relevant data about the shipment and the accessions sent into the data file on seed distribution.
2. Record the number or weight of seeds removed from each accession in the genebank management data file.

Notes and Examples

It is a good idea to keep a separate file of names and addresses of recipients and the germplasm they have received. These people should then be asked to send data on the accessions back to the genebank.

Regeneration of Accessions

What is Regeneration?

This is the renewal of an accession by taking a random sample of seeds, sowing and growing the resulting plants under conditions so that the seeds harvested will possess the same characteristics as the original population.

Why is Regeneration Necessary?

The most important part of conservation of material is that the seeds maintained in the genebank should always produce plants which show the representative characteristics of the accession. If seed viability begins to decrease, then genetic changes can occur in the seeds, thus changing the characteristics of the accession. Therefore renewal of the accession is necessary and must be carried out in the correct way to maintain the characteristics of the original population. Regeneration is also necessary when the number of seeds of the accession in store falls below a critical level.

When are Accessions Regenerated?

Regeneration is a time-consuming and expensive task which is potentially dangerous to accessions because it can allow genetic change and the accession is at the risk of pests and diseases during the growing cycle. Therefore regeneration should be kept to a minimum. It is necessary when the seeds of the accession begin to lose viability, when small seed samples are entered into the genebank or when seed stocks become low due to distribution from active collections.

The IBPGR Advisory Committee on Seed Storage recommends that for most species an accession should be regenerated when the seed viability falls below 85%. Regeneration is also required when the number of seeds in an accession falls below the number required to complete 3 regeneration cycles for that species.

If there is doubt about the condition of the seeds, their regeneration must take priority; otherwise the sample may be lost. Some very small seed samples may need several cycles of regeneration before sufficient seeds are harvested to reach the recommended sample sizes.

How should Regeneration be Carried Out?

Regeneration is carried out by growing out and harvesting a representative sample of seeds from an accession. The techniques used must ensure that the seeds produced from regeneration are fully representative of the original population.

Since species differ in agronomic requirements, breeding systems, compatibility and population structure, the techniques used are crop-specific and will not be discussed in detail in this manual. It is recommended that curators consult with plant breeders and experts on the various crops to identify appropriate methods for regeneration and that these be incorporated into the genebank's regeneration procedures.

Principle Factors to Consider When Selecting a Suitable Regeneration Method

- Suitability of the environment for each accession to ensure that natural selection is kept to a minimum and there will be good seed production.
- Special requirements needed to stimulate germination, to break dormancy and induce flowering (vernalization or photoperiod).
- Optimum number of plants to fully represent the population and ensure that all the characteristics of the accession are maintained.
- Optimum number of plants to ensure that enough seeds are produced to reach the recommended sample size.
- The correct spacing for the plants to ensure optimum seed set and production of good quality seeds.
- The breeding system of the plant (outbreeder or inbreeder) to determine whether isolation is needed.

- The need for an artificial pollination method, either by the introduction of insect pollinators or hand pollination, to ensure seed set.
- The presence of compatibility barriers to pollination which may have to be overcome artificially.
- The sexuality of the plants (dioecious, monoecious or hermaphrodite) to ensure that sufficient male and female plants are available for pollination and good seed production.
- The need to mix approximately equal proportions of seeds from different mother plants to avoid maternal effects from particular plants in the population.

Determine How Many Accessions can be Regenerated in One Cycle

1. Seek advice on the techniques and environmental conditions required for each crop from plant breeders and the curator.
2. Consider the area of the land available for regeneration.
3. Consider the number of plants per plot and their spacing to determine plot size.
4. Consider the area required for isolation and paths between plots.
5. Divide the amount of land available by the amount of land required for one accession including paths. The nearest low whole number is the number of accessions that can be grown in the space available.
6. The curator will consider the work involved and the staff available and decide how many accessions to handle.

Notes and Examples

Do not handle too much material at one time - it is better to handle fewer accessions well than many accessions badly. Generally speaking, seeds stored in the genebank can usually be kept for another year without greatly decreasing seed viability.

Decide Which Accessions to Regenerate

1. Make a list of all accessions with a seed viability of less than 85%.
2. Make a list of new accessions which contain less than the recommended number of seeds for storage in the genebank.
3. Make a list of all accessions with the number of seeds remaining less than three times the number of seeds used for regeneration of that species.
4. The curator will decide which accessions have priority. As a general rule first priority should be given to the accessions with the lowest seed

viability, second priority to any accessions which are required urgently from either list and lastly to accessions with high viability but few seeds.

5. Make a list of the accessions which the curator has decided to regenerate in this cycle and a second list of other accessions which are awaiting regeneration in the future.
6. Record the location of each accession in the genebank from the list of those to be regenerated.

Determine Where to Grow the Material

1. Check the passport data to find which environmental conditions are needed for that accession.
2. Check the number of seeds that are available.
3. If few seeds are available and the accession is sensitive to the prevailing environment, grow under controlled conditions or in a glasshouse.
4. If enough seeds are available and the accession is well adapted to the outdoor climate or special conditions are not available, grow in the field.

Notes and Examples

Select the right environment for the accession. Take special note of day-length requirements in certain crops or they may not flower. In some cases where seed dormancy is present it may be necessary to treat the seeds first before sowing in the field.

Prepare the Planting Plan

1. Draw a scale map of the area to be used.
2. Mark this plan off into plots of the correct size for each species, allowing room for isolation blocks, inter-planting of species and paths for access.
3. Fill in the accession numbers from the list already prepared into the appropriate plots, taking account of any isolation distances, species and maturity differences necessary to prevent crossing between plots.

Prepare the Seeds for Planting

1. Remove the container from the genebank.
2. Allow the container to warm to room temperature before opening. This may be several hours for large containers.
3. Write the accession number and genus and species of the seeds on the outside of a seed packet.
4. Open the container and quickly weigh or count the required number of seeds already decided as being appropriate for the regeneration of

accessions of that species, but do not use more than half of the seeds which are available for that accession. Work in a dehumidified room, if possible and close the containers as soon as the seeds are removed to prevent the seeds absorbing moisture from the air.

5. Place the seeds in the labelled paper packet and seal.
6. Record the number or weight of seeds removed from each accession.

Notes and Examples

Always use random sampling techniques to remove seeds from the container so that the resulting sample will provide a good representation of the original population.

Proceed with the Regeneration

1. Follow the planting plan using techniques outlined by the breeders and curator to ensure that regeneration is done well and no crossing occurs between accessions.

Harvest the New Seeds

1. Harvest each accession separately when the seeds are fully mature.
2. Mix the same weight of seeds from all the plants of one accession to minimise maternal effects and still keep as much variation as possible within the sample. Make sure that population samples are not divided into lines but keep the accession complete.

Notes and Examples

Make sure that the population is kept whole and not split into lines. If lines are required for plant breeding needs, it is possible to keep these separate in a different programme, but the population should also be maintained intact.

Process the Seeds and Enter New Seeds into the Genebank

1. Carry out seed cleaning (Section II) and measure the weight or number of true seeds harvested.
2. Do not mix the seeds from the different regeneration cycles.
3. Use the seeds from earlier regeneration cycles to supply seed requests.
4. Regeneration may need to be carried out several times to reach the recommended sample size for genebanks (Section VII).
5. Use the other half of the original sample for the second regeneration and representative samples from these two for further regeneration.

9

Management of Seed-Borne Diseases

Measuring the Amount of Seed-Borne Inoculum

The amount of inoculum may be expressed in terms of the proportion of infected seeds, the degree or severity of infection (inoculum per individual seed) or the viability of the inoculum (i.e. the infectivity of the pathogen in seed). Most seed tests measure the proportion of infected seed. Because low levels of seed-borne inoculum can lead to considerable disease, the most sensitive test should be used to determine the level of seed infection.

Fungi may be detected using a standard blotter test or an agar plate test, the latter being the more sensitive. Seed-borne bacteria can also be detected using an agar plate test. While results of standard tests indicate the proportion of infected seed, they provide no information on the amount of inoculum per seed. Where a high percentage of seed is infected there is often more inoculum per seed associated with larger infections and deeper penetration. Seed-borne viruses are usually detected using ELISA or PCR tests. It is important that the diagnostic tests are conducted on germinated seed (seedlings) as virus may sometimes infect the seed testa without infecting the embryo or seedling e.g. PSbMV.

There is a wide range of tolerance levels for different pathogens. For viral diseases a threshold of <0.1% seed infection is recommended for sowing in high risk areas, and <0.5% seed infection for sowing in low risk areas. For most fungal pathogens a threshold of <1% seed infection is acceptable. However, there is a nil tolerance for the most serious fungal diseases e.g. Ascochyta rabiei in chickpea.

Preventing the Spread of Seed-Borne Diseases

The best method of to reduce the risk of disease damage is to source pathogen-free seed. Testing seed before sowing will establish whether seed is free of disease. The next best option is to select seed from crops that show no sign of disease. Seed with high levels of seed-borne disease should not be used for sowing. For some fungal diseases, it may be possible to reduce the risk of disease by applying a fungicide to seed prior to sowing. However, seed treatments are not available for the control of virus or bacterial diseases.

Seed Testing Laboratories

1. AgriFood Technology Box 728, Werribee, Vic 3030. Telephone: 03 9742 0555, Fax: 03 9742 4228.
2. AGWEST Plant Laboratories Department of Agriculture Western Australia, 3 Baron Hay Court, South Perth, WA 6151. Telephone: 08 9368 3721, Fax: 08 9474 2658.
3. AsureQuality 3-5 Lillee Crescent (PO Box 1335) Tullamarine Vic 3043 Telephone: 03 8318 9024
4. SARDI Seed Testing Service, GPO Box 379, Adelaide, SA 5001. Telephone: 08 8303 0360, Fax: 08 8303 9384.
5. TASAG ELISA and Pathology Testing Service Department of Primary Industries, Water and Environment, 13 St John's Avenue, New Town, Tasmania 7008. Telephone: 03 6233 6845, Fax: 03 6278 2716. Note: Please check with the TASAG laboratory if the test is required to satisfy phytosanitary or export purposes.

Factors Affecting Forage Stand Establishment

The establishment of a uniform and vigorous stand is of paramount importance to the successful production of most crops. Without a good stand, the effectiveness of other agronomic inputs is drastically reduced, and usually such inputs can never compensate for the negative impact of a poor stand. With perennial forage crops, failure to achieve a dense stand at initial planting can have long-lasting consequences compared with annual crops. This is especially true for seeded annual and perennial forages that exist as discrete plants (e.g., alfalfa [*Medicago sativa* L.] and lovegrass [*Eragrostis* spp.]) (Miller & Stritzke, 1995). Obtaining dense stands from initial plantings of forages which vegetatively propagate (e.g., through rhizomes or stolons) is less critical; however, establishment of healthy plants that are able to compete and withstand abiotic and biotic stress is still very important with those species. This paper focuses on aspects of seed quality, seed enhancement, and recent developments in stand establishment practices of temperate forage crops grown in crop rotations in the USA (e.g., perennial grasses and legumes grown for hay or short-term pastures). The improvement of long-lay permanent pastures and rangelands through artificially enhanced and natural processes of seedling recruitment of desirable species is beyond the scope of this paper.

Seed Quality

The importance of quality seed to successful forage establishment cannot be minimised. Several authors have recently reviewed aspects of forage seed development, processing, germination, and quality (Hill et al., 1997; McDonald et al., 1996; Simon et al., 1997). The forage seed industry in the USA delivers good quality seed to the market place; however, from a practical stand establishment perspective, there is a need to improve and more widely utilise methods of assessing forage seed vigour.

The standard germination test does not relate well to actual field emergence (Assoc. Off. Seed Anal., 1983), particularly when emergence conditions in the field deviate from optimal. Seed lot vigour becomes very important when environmental stresses occur at sowing (Hill et al., 1997). Forage seed lots with high germinability often differ in vigour, resulting in reduced establishment, dry matter production, and performance in storage (Hampton, 1991). Seed vigour may also influence nodulation and dinitrogen fixation in forage legumes, as in field beans (*Phaseolus vulgaris* L.) (Rodriguez & McDonald, 1989). An effective assessment of seed vigour would be useful for positioning of seed lots in the market place. For example, seed vigour tests might help identify which seed lots would perform best under the cold and wet spring planting season as opposed to the warm and drier late summer planting season.

Research on vigour testing of forage seed lots has received little attention compared with that conducted on other species. Several researchers have evaluated the use of accelerated aging and the conductivity test for forage seeds (Clark, 1982; Hall & Wiesner, 1990; Hampton & Bell, 1989; Helmer et al., 1962; Hampton & TeKrony, 1995; Wang and Hampton, 1993). Standardized procedures for these tests have been developed for several forage species, and hopefully their use will increase and assist the industry in assessing seed lot planting value of seeds for different markets. Hall & Wiesner (1990) concluded that a combination of tests which include both physiological and biochemical measures of seed quality should be used in adequately assessing seed vigour. This promises to be a fruitful area of research with potential to provide real and practical benefits to forage producers.

Seed dormancy in forage grasses and hard seed content in forage legumes are two important factors affecting the stand establishment. In nature, these mechanisms ensure a broad, dormant soil seed bank capable of germinating when favourable conditions are encountered for growth and establishment

(McDonald et al., 1996). They also ensure that not all seeds germinate solely on a single environmental cue, but that continued emergence and establishment of the species occurs under a variety of habitats and environments (McDonald et al., 1996).

Although seed dormancy is advantageous in nature, production agriculture demands that seeds planted germinate uniformly and quickly to produce a productive stand. Therefore, various mechanisms are employed to break seed dormancy. McDonald et al. (1996) described these methods for grasses, including KNO_3, light treatments, and prechilling. Various methods can be used to decrease hard-seededness in forage legumes (Bass et al., 1988), but mechanical scarification is probably the most commonly used in the industry. Mechanical scarification injures seeds, thus resulting in higher abnormal seedling counts and deterioration of vigour and viability. A recent study demonstrates that scarification may not always be necessary to ensure adequate field performance of forage legume seed lots with high hard seed content (Albrecht et al., 1996). Alfalfa seed lots with hard seed levels of up to 50% had no significant impact on total seedling emergence, establishment year forage yield, or forage yield the year after establishment. Rate of emergence was reduced slightly by high levels of hard seed. Scarification increased rate of emergence of most seed lots, but had no effect on forage yield. These results indicate that it may not be necessary to mechanically scarify alfalfa seed lots with moderate levels of hard seed content in order to ensure good performance in the field. There is a need for further study on factors governing hard seed expression under natural field conditions, and methods developed which adequately predict field germination and emergence of forage legume seeds.

Seed Enhancements

Various seed enhancements have been evaluated for improving planting value and quality of forage seeds. These include various seed treatments, seed coating, pelleting, and controlled hydration (refer to McDonald et al., 1996; Johnson, 1975). Unfortunately, the high cost of some of these enhancements has limited their commercial use in forage seeds. Forage seeds have a lower expectation of return compared with vegetable and flower seeds; thus the additional cost of any seed enhancement must be low in order for a profit to be realised. Seed treatment and seed coating have been the most widely used enhancements in the forage seed trade, as evidenced by the amount of commercial investment into these processes (Ni et al., 1998; Simon et al., 1997). As the cost of other types of enhancements is reduced, their use in forage seeds will likely increase. Seed treatment in forages involves primarily the application of fungicides or insecticides, and bacterial inoculants in the case of nitrogen-fixing legumes. Seed treatment is now a very common practice in the forage seed industry in the USA, particularly with alfalfa seed. Seed coating involves the addition of either inoculant, fungicides or insecticides together with other substances to provide a protective barrier over the seed. The current coating process in the forage seed industry employs limestone coating and a bacterial inoculant in the case of legume seed (Ni et al., 1998). A systemic fungicide is often added in the coating process, usually metalaxyl [*N*-(2,6-dimethylphenyl)-*N*-(methoxyacetyl)alanine methyl ester] for the control of *Pythium* and *Phytophthora* spp. Nutrient elements may also be included in some cases. The normal coating percentage is 33% for small-seeded legumes (i.e., seed:lime ratio=2:1) and 50% for grass seeds (i.e., seed:lime ratio=1:1). Usually all the ingredients are added to the seeds in one step.

Seed coating of alfalfa has been shown to improve *Rhizobium* survival and early nodulation of seedlings (Horikawa & Ohtsuka, 1995a, 1995b; Sheaffer et al., 1988). Coated alfalfa seed has not shown a consistent advantage over uncoated seed in terms of higher stand densities and forage yields. Some studies have demonstrated significant improvements in stand establishment and forage yield while others have demonstrated mixed results across different environments (Canestrino et al., 1998; Horikawa et al., 1996; Twidwell & Gallenberg, 1993; Sheaffer et al.,1988; Marble et al., 1990). The decision on whether to use an alfalfa seed enhancement should be based on soil type, field history of disease problems, and cost of the treatment (Twidwell & Gallenberg, 1993). Johnson (1975) indicated that further research is still required to determine the specific situations in which coating and pelleting may be of value in improving rangelands. Methods of reducing the cost of seed enhancements are needed so the industry can justify its cost on seeds having lower margins of return.

Recent Developments in Forage Stand Establishment Techniques

Considerable research has been conducted on stand establishment of forage crops. There are numerous effective methods of establishing forage crops, but all employ several key principles which ensure a favourable environment for seed germination and seedling establishment. These include proper seed placement, good seed-soil contact, proper time of planting, adequate seeding rate, minimising competition from other species (e.g., weeds or companion crops), and protection from insect pests and diseases. A combination of cultural practices and genetic improvement has contributed to successful stand establishment.

In practice, poor seed placement in the soil is probably the most frequent factor leading to establishment problems. In such cases, the seed is either planted too deeply or adequate coverage and good seed-soil contact are not provided. These problems usually occur because of poor seedbed preparation and/or inadequate adjustment of planting equipment. Recent advances in planting equipment have focused primarily on planting of large-seeded row crops. In some cases, these "improved" planters are less suited to planting small-seeded crops than the older equipment. The advent of no-tillage planters has resulted in an increased interest in reduced tillage and no-tillage seeding of forages. Forage establishment using no-tillage techniques is not a new practice, having been used in renovating pastures. But the wider use of no-till planting equipment in row crops has increased the opportunity for no-till seeding for establishing new forage stands. No-till seeding presents important challenges in stand establishment.

Seed-soil contact is probably the biggest challenge in no-till systems, especially since most temperate forage species have very small seeds. Further improvements in no-till planting equipment are necessary to achieve consistent success with forage seeds. Most no-till planters were designed for larger-seeded crops. The Truax drill is one example of a no-till drill designed specifically for grassland species, with the ability to handle very diverse seed types. It has not been widely adopted in regions of the USA where forages are grown in rotations with row crops because of its high cost and lack of adaptability to seeding large row crop species.

No-tillage seeding of forage legumes has also increased the incidence of Sclerotinia crown and stem rot, one of the most destructive diseases of forage legumes in the eastern and southern USA. This fungal disease can completely destroy newly established stands, and is more prevalent in late summer seedings which are established using no-tillage and reduced tillage methods. Breeding for resistance to this disease and management practices to reduce its incidence are in progress (Sulc & Rhodes, 1997).

Modified grain drills and cultipacker seeders are considered the most reliable seeding implements for small forage seeds; however, some producers consider them expensive and too slow for seeding large areas. The old seeding technology of broadcasting seed has recently been modified by innovative producers seeking to increase the efficiency of forage seeding (Barnhart, 1998).

Commercial and farmer-owned dry fertilizer spreaders, which are capable of delivering material over a very wide swath, are used to rapidly broadcast a blended mixture of dry granular fertilizer and seed. The most modern dry fertilizer applicator equipment uses flowing air to move the dry fertilizer material through a manifold and a series of pipes to provide a more uniform distribution of material than is possible with the older "spinner" spreaders. This technology requires well-prepared and firm seedbeds before broadcast spreading, followed by the use of cultipackers (rollers) to provide shallow seed coverage with the desired seed-soil contact. Forage legumes and cereal seeds are the most common species seeded with this method.

With the resurgence and improvements in broadcasting seed with dry fertilizer, the development of broadcast sowing of legume seeds with fluid fertilizers soon followed. A significant portion of the fertilizer applicator industry has adopted the sale and distribution of fertilizer in the fluid form. The fluid fertilizer applicators have the ability to broadcast free-flowing liquids or finely divided fertilizer material suspended in water, often with a clay-based suspension carrier. A variety of materials can be blended and accurately metred across a wide swath (18 metres). Through trial and error, fluid fertilizer applicators have perfected an efficient fluid seeding system: all screens are removed, 2 to 3% clay is used with a minimum of 470 to 560 L ha-1 of fluid fertilizer (or 3 to 4% clay is used with water), the clay/fluid suspension is agitated for 0.5 hr, and the desired amount of legume seed is added to the agitating suspension in the tank, and then applied immediately to the field. Large flood-jet nozzles are required. As with dry fertilizer seeding, conventional tillage must be used to prepare a clean, firm seedbed before application, and the field must be cultipacked after the seed is applied to provide shallow seed coverage and good seed-soil contact.

Seeding legume seeds with fertilizer led to concerns of a detrimental "salt effect" of fertilizer on alfalfa seed germination, seedling vigour, and *Rhizobium* inoculant survival. Limited greenhouse studies (Evers, 1986) demonstrated that seed germination and seedling vigour of forage legumes began to decline after 12 to 18 hr when in contact with granular fertilizer. The fate of *Rhizobia* on the seed or nodulation was not monitored in those studies. Johnson (1997) evaluated performance of alfalfa seed in fluid fertilizer suspensions. From bacterial culture, controlled environment, and field studies, he concluded that *Rhizobium* death and alfalfa seed emergence with use of 4-10-10 and 10-34-0 fluid fertilizers as carriers should not be a concern to the producer. Seeding alfalfa with fluid fertilizers should be considered if timeliness of operation can be improved and corrective nutrient needs met with the fertilizer application.

Although advances have been made in our knowledge of forage seed physiology, technology, and stand establishment practices, there is much room for improvement. Stand establishment continues to be one of the most common production problems of forage crops in the USA. We have mastered the establishment of forages crops under good conditions, but forages are often grown on marginal and less productive soils than row crops. Abiotic and biotic stresses are exacerbated on these less productive soils. There is a need to more thoroughly understand the factors affecting forage stand establishment under biotic and abiotic stress so that means of improving establishment success can be developed. Undoubtedly improvements in stand establishment will result from advancements in seed quality and seed enhancements, genetic improvement, as well as improved seeding techniques.

Hybrid Rice Seed

The success and sustenance of hybrid rice technology depends on the efficient and economic hybrid *seed production,* so as to provide good quality seed at reasonable price to the farmers. Hybrid seed production, in self pollinated crop like rice, is quite tedious. Therefore, extensive trials were conducted to standardize the seed production package. Various requirements for successful seed production such as ideal locations/seasons, seed rate, row ratio, GA3, supplementary pollination techniques etc., have been worked out and a seed production package has been developed. The details of the package are as under.

Optimum package for hybrid rice seed production and CMS multiplication

Activity		*Particulars*
Seed rate	:	Seed parent 15 kg/ha Pollen parent 5 kg/ha
Nursery	:	Sparse seeding to ensure multitillered (4-5) seedlings in 25 days
Row ratio	:	2B:10A, for CMS multiplication 2R:12A, for hybrid seed production
Number of seedlings/hill	:	2 seedlings/hill for seed parent; and 3 seedlings/hill for pollen parent
Spacing	:	B/R to B/R - 30 cm B/R to A - 20 cm A to A - 15 cm
GA_3 application	:	45 g/ha at 5% heading in two split doses on consecutive days
Supplementary pollination	:	Twice a day at peak anthesis during flowering phase
Roguing	:	Twice during vegetative phase based on morphological characters and twice during and after flowering based on floral characters sterility etc.
Seed yield (t/ha)	:	1.0-1.5

Seed producers can obtain seed yields of about 1.0 to 1.5 t/ha by adopting this package. There is a need to refine the present technology so as to enhance the seed yields up to 2.0 to 2.5 t/ha so that the seed cost can be reduced.

GA3 is one of the essential requirements for hybrid seed production and it is a costly input. Therefore, efforts are being made to find some alternate chemicals or substitutes to economize GA3 use. Initial results have indicated that GA3 requirement can be reduced by using some chemicals such as ZnSO4, urea and boric acid in combination with lower doses of GA3. Such studies will help to reduce the cost of seed production thus helping the farmers to get the hybrid seed at reasonable cost which is now being sold @ Rs. 90-100/kg.

India has a very well organized seed industry both in public and private sectors and this can be effectively utilised for hybrid rice seed production. Realising fully well, the potential of seed industry, enough encouragement was given to the seed sectors to take up hybrid seed production. Required parental lines were made available to the interested companies and their personnel were imparted training on hybrid rice seed production technology. This step has really helped many companies and some of them have established full fledged R & D facilities for hybrid rice. Now the seed industry has fully geared to meet the seed requirement and what it needs is the demand which has to be created by well organized transfer of technology activities.

Large Scale Seed Production

Large scale seed production has been taken up predominantly by private sector and to lesser extent by public sector seed agencies in India during the last five years. Initially seed yields obtained were less than 1.0 t/ha but now an average seed yields of 1.5 t/ha or more are being obtained routinely. Maximum seed yield of 3.8 t/ha has been recorded in a seed production plot of 2.0 ha area by a private sector company.

Progress made in large scale seed production in India, in the initial phase, has been impressive as compared to the progress made in China at similar stage. In China, it took more than five years to attain an average seed yield of 1.0 t/ha and only after a decade average seed yields of 1.5 t/ha were attained. In India, average seed yields of 1.5 t/ha are being obtained after five years of initiating large scale seed production. Seed yields can be increased further by utilising the new CMS lines with higher outcrossing potential, now under development in Indian hybrid rice breeding programmes and through further refinement of seed production technology.

Now 3,000-3,5000 tons of hybrid rice seed is produced annually in India, 95% of it by the private seed sector. The leading private sector seed companies in Hybrid Rice Seed Production are Hybrid Rice International, PHI, Mahyco, Indo-American, Parry Monsanto, Advanta India, Syngenta and Nath Seed company. Half a dozen other private seed companies are also engaged in hybrid rice seed production on a smaller scale.

Among public sector seed agencies, National Seed Corporation (NSC), State Farms Corporation of India (SFCI), Andhra Pradesh State Seed Development Corporation (APSSDC), Karnataka State Seed Development Corporation (KSSDC), West Bengal State Seed Development Corporation (WBSSDC), and U.P. and Terai Seed Development Corporation (UPTSDC) are presently engaged in hybrid rice seed production on much smaller scale. All these public sector seed agencies put together, produce around 100 tons of hybrid rice seed annually.

Hybrid rice is likely to become popular and adopted in large acreage during the next decade. Huge quantities of hybrid rice seed will be needed. Thus there is ample scope for all the private and public sector seed agencies in India to produce large quantities of hybrid seed in the years ahead to popularize rice hybrids.

Hybrid Seed

The agriculture and gardening, hybrid seed is seed produced by cross-pollinated plants. In hybrid seed production, the crosses are specific and controlled. The advantage of growing hybrid seed compared to inbred lines comes from heterosis. To produce hybrid seed, elite inbred varieties with well documented and consistent phenotypes (such as yield) are crossed and the resulting hybrid seed is collected. Another factor that is important in hybrid seed production is the combining ability of the parent plants. Although two elite inbred parent plant varieties may produce the highest yields of their crop, it does not necessarily mean that crossing these inbreds will result in the highest yielding hybrid. Combining ability is the term used to describe the level of heterosis that the parents will generate in the resultant seed. Higher combining ability between the parents results in increased performance in the resulting hybrid seed. Hybrids are bred to improve the characteristics of the resulting plants, such as better yield, greater uniformity, improved colour, disease resistance, and so forth. Today, hybrid seed production is predominant in agriculture and home gardening, and is one of the main contributing factors to the dramatic rise in agricultural output during the last half of the 20th century. In the US, the commercial market was launched in the 1920s, with the first hybrid maize. All of the hybrid seeds planted by the farmer will be the same hybrid, which causes the first generation of seed from the hybrids planted to be inbred. This is why hybrid seed is generally not saved from subsequent generations and is purchased for each planting. Hybrid seeds are much dearer than normal seeds, due to the technology, time and effort put in to produce them.

Bibliography

Alfred Steferud: *Diseases of Fruits and Nuts*, Biotech Books, Delhi, 2005.

Bandyopadhyay, P C: *Breeding and Crop Production*, Gene Tech Books, Delhi, 2007.

Cullis, A. : *Rainwater Harvesting: The Collection of Rainfall and Runoff in Rural Areas,* London, U.K.: IT Publications, 1986.

Dabholkar, A.R.: *General Plant Breeding*, Concept, Delhi, 2006.

Dirr, Michael A.: *Manual of Woody Landscape Plants,* Stipes Publishing, L.L.C., Champaign, 1998.

Doijode S. D.: *Seed Germination in Fruits*, New Delhi, Malhotra Publishers, 1993.

Engelman, R., and P. LeRoy: *Sustaining Water Washington*, D.C.: Population Services International, New York, 1993.

Erik Nissen-Peterson: *Rainwater Catchment Systems*, UK: Intermediate Technology Publications, 1999.

Galili G, Kigel J.: *Seed Development and Germination*, M. Dekker, New York, 1995.

Gliessman, Stephen R.: *Agroecology: The Ecology of Sustainable Food Systems*, CRC Press, Boca Raton, 2007.

Gour, H.N.: *Integrated Plant Pathology*, Scientific, Delhi, 2009.

Graf, Alfred Byrd: *Advances in Plant Physiology*, Rajat Pub, Delhi, 2008.

Gurjar, Ram Kumar: *Geography of Water Resources*, Rawat, Delhi, 2008.

Hartmann, Hudson Thomas, and Dale E. Kester: *Plant Propagation Principles and Practices*, Prentice-Hall, Englewood Cliffs, N.J., 1983.

Hashimoto, Y.,: *The Computerized Greenhouse: Automatic Control Application in Plant Production*. San Diego, Calif.: Academic Press, 1993.

Herminie Broedel Kitchen: *Soils and Crops: Diagnostic Techniques*, Satish Serial Publishing, Allahabad, 2004.

Jacquat Christiane: *Plants from the Markets of Thailand*, Bangkok, Duang Kamol, 1990.

Jeet Inder: *Rainwater Harvesting*, Mittal, Delhi, 2009.

Jones, R. M.: *Plant Resources of South-East Asia,* Wageningen, Pudoc Scientific Publishers, 1992.

Kataria, T N: *Plant and Crop Physiology*, Pearl Books, Delhi, 2008.

Macself, R: *Soils and Fertilizers*, Satish Serial Pub, Delhi, 2005.

Morhardt, Sia; Morhardt, Emil; Emil Morhardt, J.: *California Desert Flowers: An Introduction to Families, Genera, and Species*, University of California Press, Berkeley, 2004.

Patel, GM: *Botanical Pesticides for Pest Management*, Scientific, Delhi, 2008.

Pilgrim R.: *Post Harvest Handling of Minor Exotics*, St. George's, Grenada, 1996.

Rost, Thomas L.; Weier, T. Elliot; Weier, Thomas Elliot: *Botany: a Brief introduction to Plant Biology*, Wiley, New York, 1979.

Sambamurty, A V S S: *Textbook of Plant Pathology*, I K Pub, Delhi, 2006.

Smith, Welby R.: *Orchids of Minnesota*, University of Minnesota Press, Minneapolis, 1993.

Stern, Kingsley R.: *Introductory Plant Biology*, Wm. C. Brown Publishers, Dubuque, IA, 1991.

Taylor EL, Taylor TMC: *The Biology and Evolution of Fossil Plants*, Prentice Hall, Englewood Cliffs, N.J, 1993.

Thapliyal, B K: *Democratisation of Water*, Serials Pub, Delhi, 2008.

Thind, T.S.: *Annual Review of Plant Pathology*, Scientific, Delhi, 2004.

Thomas E.: *Fruit Production in St. Kitts and Nevis*, Port of Spain, IICA, 1996.

Verma J.P.: *Detection of Plant Pathogens and their Management*, Angkor, Delhi, 1995.

Walton, William C.: *Groundwater Resource Evaluation*. McGraw Hill, New York, 1970.

Whealy K.: *The Garden Seed Inventory*, Decorah, Seed Saver Publications, 1988.

Willard, C. J.: *An Experimental Study of Sweetclover*, Ohio Agricultural Station, Wooster, Ohio, 1927.

Index

❑❑❑